The Story of Everything

The Double Existence Model

The Story of Everything
The Double Existence Model

Author	: Vadi Dipçin
Editor	: Burçe Kaya
Cover Image	: Göktürk Dipçin

X	: @dblexistence999
Instagram	: @doubleexistence999
E-Mail	: doubleexistence999@gmail.com

September 2023

KDP-ISBN: 9798859768523

Thanks and dedicated to,

all the scientists who have worked to enlighten us with the light of truth despite all the obstacles and difficulties throughout history.

Vadi Dipçin was born in 1971 in Istanbul. He received his BS, MS and PhD degrees from Istanbul Technical University, Electronics and Communication Engineering Department. He did academic studies on digital TV systems, image and video processing. After working as a research assistant at the University's Video Education Center for two and a half years, he moved to the private sector. In his 25-year career, he held strategy, business development and product management related positions in the consumer electronics, telecommunications, media industries. He worked in technology and service-based projects such as TV platforms, fixed and mobile broadband infrastructures, and content production. Dipçin, whose favorite hobby is physics, is also an amateur short film director.

Table of Contents

Foreword

If you'd categorize this book, you probably would have labeled it as a popular science book. It is about physics, and it contains many scientific information and ideas. You need to have a background about subjects such as Relativity and Quantum in detail to be able to understand the content entirely.

On the other hand, this book is about understanding our reality. Physics is the core of all other natural sciences, and it tells us how and why everything is as it is, gives us clues about the nature of existence. To understand physics at this level is to understand our existence. Thus, the outcomes in this book are for everyone who is curious about it.

That's why it was hard to write this book. I have thought how to make it possible to write a physics book with new ideas at the edge of science, almost a scientific paper, which must be understood by anyone for a very long time.

The book has two purposes:
1. To present a new scientific model
2. To answer questions about our existence and reality on top of this model

So, I decided to arrange the chapters in a way which could serve for both purposes and that any reader segment can enjoy it.

If the scientific and mathematical details are not interesting for you, you can skip Section 3.5, Chapter 4 and or 5 partially, or even completely. These chapters serve item 1 above. Chapter 6 contains the core ideas in Chapter 3, 4, 5 in plain

English with additional comments which will help you understand the context and the Model's outcomes about our existence.

If you are interested in the full scientific scope, you can read the book as it is. I still recommend you to read Chapter 6. Chapter 4&5 contain all the hypotheses. Chapter 6 contains a subset of them which will complete the story about reality and comments about it. So, there is also some additional content in Chapter 6.

There are many references and additional information I could include in the book to support the ideas about DEM, but I didn't. I preferred to explain the core ideas in the most basic way, to be able to clarify them briefly. Otherwise, the book would become a huge scientific paper and not be readable as a book. I know that the physicist who'll read this book will make those connections easily.

The Double Existence Model and the content of this book are brain storming sessions about the universe and the facts of our reality which I hope, will help us all to understand the truth further.

Vadi Dipçin

1- Questions About the Universe

It was 1977 when I have first read an article about physics and astronomy first time in my life. I was six years old and wasn't even going to the primary school yet. That morning, I started to read the usual daily newspaper at home and realized that there was a new Kids Magazine attached. That was a good surprise for me, so I started to read it. I saw a headline on one of the pages. The headline was something like "The Sun is going to die in 50 billion months". I was not only surprised but also disappointed. The Sun had an end. I had no idea about that before. I have read the article and tried to understand what was going on. I don't remember the text, but it was sure about nuclear fusion, the amount of hydrogen in the Sun etc. After some time, when I have felt better, I have divided 50 billion by 12 to convert months to years and it was clear to me that this was not a danger, I would face. It was a relief, and I skipped the page.

70's and 80's were the years of so called "Space Age". 21st Century was coming up and there were big expectations about it. Science fiction was an important genre and many legendary TV series, and movie titles were released. Star Trek, Star Wars, Battlestar Galactica were in my favorites list. Space and science meant the world to me. When I was in primary school, my dream was to be an astronaut. I wanted to go to space. This was a very clear target for me. My plan was to be very rich, design and manufacture my own spaceship[1]. I was somehow sure that I could do that.

---

[1] So, you can guess, I do appreciate and admire Elon Musk. He made what I dreamt about. The funny thing is my dream at high school was to design and manufacture electric cars. I was working on an idea about a very efficient electrical engine. You got what I mean ☺

That's why my mother made something which changed my life. Probably, I would go to the same direction anyway but that was the trigger. It was June 1982, summer holiday, right after I graduated from the primary school. She saw a science magazine in a bookstore and thought I'd love to read it and bought one. The name of the magazine was "Bilim Dergisi"[2]. Charles Darwin was on the cover, holding the hand of a monkey. This was the turning point in my life. Thanks to this magazine, I met with Isaac Asimov, Arthur C. Clarke, Carl Sagan and other important authors and I started to read science fiction and science books, especially about physics and astronomy.

It was October 1982 Issue where I read an article about relativity, quantum, unified fields, Einstein, and Bohr, first time in my life. That's where I saw the light. I was only 11 years old, had no idea about calculus, differential equations, or probability distribution functions but I could understand the ideas like time passing differently at different speeds. I had many questions in my mind about time travel and the origin of the universe. I knew I had to learn physics to get some answers. Physics is the only possible source which can answer questions like these. All other sciences are just subsets of physics and physics is the only science which directly studies the "origin".

The scope of my curiosity was clear. I just wanted to look at the universe from outside of it to see "HOW" things were working. It was the most important question to me. Of course, I was curious about the other important questions as well: "WHY?". Why do we exist? Why is everything like this?". These are the most difficult ones and their answers, if any, can only

---

[2] "Science Magazine" in English

be found answering the "how" questions. That's why I was focused on understanding how the universe was structured.

The main big question in my mind was: "Should anything exist at all?". This is a very difficult concept to think about. Make the matter, energy, space-time, the universe, and anything beyond it disappear and go on like this. What does remain? Should anything remain? Why does anything exist?

Thus, as you can see, the focus goes "beyond the universe" or "outside the universe", name it as you like. All my questions were a part of this concept. "What is the time, how many spatial dimensions are there, how big is the universe?"

I had a very clear answer to the last question when I was like 16-17 years old. I have already built a universe model myself, using all the information I have read until then. The universe should have zero volume when we looked at it from outside, so an infinitely small, punctual structure, a singularity including the time as an additional dimension. All the 4 Space-time dimensions were of zero length when you looked from outside of the universe. Everything was "at the same place" and was happening "at the same time" point. The volume of the space and the time we experience were internal attributes of the universe.

The idea about spatial dimension being a singularity was obviously false, that's why I'll skip that part, but the time version of the idea is an interesting starting point to think about our reality.

1.1- Everything happening "at the same time"

This was the second component of my model about the universe. The past, present, and future **exist** all "at the same time" or "simultaneously" but we experience them in an order. Of course, this is a widely discussed idea. I read many discussions about it later. It was good to know, at least this section of my model was part of the scientific discussions.

The idea can be explained by using an analogy a of roll of film which has all the frames of a movie. All the objects, people and events are recorded on the film and they constitute the "movie universe"[3]. The roll is in a projector, ready for the next show. Whenever the show starts, the projector starts to project each frame in the roll to the white screen.

Each frame appears on the screen according to the sequence in the roll. The audience in the theater experiences them in that order. This is also true for the characters living in the movie universe. Neither the viewers nor characters in the movie know what's next in the movie universe, so the future is uncertain for them. Let's take an example. There is a war scene on the white screen which means "there is a war NOW". At that moment nobody knows whether the character in the war scene is going to die or survive. However, the future is already printed on the roll. All the scenes, the ones the viewer already watched and will watch, are there. The character is

---

[3] I'll use the concept of the "movie universe" along the book. In this analogy, the movie on the white screen represents our universe and we, as the viewers, are the external observers of the universe. You can think, the white screen kind of represents spacetime where events take place and time passes with each new frame.

already dead or has survived in 2000 frames. It has not been projected to the white screen yet.

Let's analyze the experience in the eyes of the audience in the theater, outside of the "movie universe" and the characters in the "movie universe".

First Scenario

The audience watches the movie looking at the white screen, so, he and the characters have the same time experience in and out of the movie universe. "The present" for the audience and for the characters is the frame on the white screen. The projector is a kind of time machine which transports the viewer and the characters to the next frame i.e., next moment, forward in time. They don't know anything about the future. Everything happens in an order.

Second Scenario

In fact, each moment is already there on the film roll. You must obey time if you watch the movie on the white screen but if you could look at the roll in the projector, you could access any moment of the movie arbitrarily. So, if the audience can look and see every frame in the roll at once while the movie is playing on the white screen, then he/she can "experience" any moment of the movie universe. He/she can jump to any random frame (time travel in the movie universe) or if he/she can comprehend all the frames all together, he/she can even experience all the events in the movie at once.

The audience would know whether the character dies or survives even before the war starts in the "movie universe". On the other hand, in the "movie universe" where the characters live, everything is still happening in an order. The characters don't even know that there will be a war, at the beginning of the movie.

In this case, the events occur in the "movie universe" in time, but the external observer, not bounded/limited with the time flow in the "movie universe", can observe all the events at once. For this observer, these events, sequenced in time in the movie universe, are time independent and can be equally accessed/experienced.

The more interesting scenario is the second one because the time experience of the audience and the movie characters diverge. The events don't occur at the same time according to the refence of the movie characters. Though an external observer can observe or experience every event in the "movie universe" at the same time, like "everything happening at the same time", which means timeless.

Using this analogy, the notions like the past, the future and the order of events are just perception of us living in the movie universe. If one can go beyond it, as an external observer outside of the universe, he/she can observe the past, present, and future "at the same time[4]" or using a more correct expression "independent of time". So, beyond the universe, the time dimension becomes zero, obsolete and meaningless...

---

[4] There is no time passing for him/her, as we know. So, it is not correct to say that "he/she observes everything at the same time". This is just an expression to describe the situation so that everyone can easily imagine what are we talking about.

1.2- High Frequency Universe

Another component in my universe model was about its fundamental way of existence. The question is: "Does the universe exist continuously or discretely?"

What does it mean? If we ask this question in a simpler and detailed way: "Is time continuous or discrete? Do we exist all the time or do we disappear and reappear again and again very quickly so that we even don't feel?"

Let's go back to the "movie universe" analogy. There are different standards for different type of cameras and display systems (like TV and Motion Picture cameras). Let's take the classic Motion Picture cameras. In these cameras, 24 Frames are taken every second. This is the original industry standard for movies.

The human eye has a weakness which makes recording and displaying of videos possible. It's the limited sensitivity to fast motion. The human eye cannot distinguish separate frames as standalone pictures if they change more than 16-17 times in a second. We perceive them as a playing video.

If more frames are displayed in one second, the quality of the video is enhanced. Therefore, 24Hz is chosen historically to record movies. If you read the specifications of your mobile phone's camera, you'll read expressions like "Full HD, 60p recording". Your phone can record and display 60 frames per second so that your video experience is even smoother than 24Hz recording. In 60Hz, fast motion can be captured and displayed more clearly. It looks more realistic with less disturbing artifacts.

What if the universe is constructed in a similar way? What if the universe is flashing very rapidly, one million times, one billion times or at a much bigger rate per second and we have the experience of continuity?

If the universe also "flashes" like this, then its existence must be related to a kind of energy wave with some frequency. Let's brainstorm on this idea, briefly.

First, if the existence of the universe is not continuous and flashing/reappearing at a frequency, then it must a very high frequency. At least, it must be higher than the frequencies of observable energy waves. Otherwise, the universe couldn't contain those energy waves. Why is that?

Think about the movie universe. Let's say our recording is only one frame per minute which also means we display one frame per minute. We shoot the display of a digital watch. In this case, the digital watch would display the change in every second, but the camera would miss all of them and would only show the changes in minutes. So, it cannot represent the change in time in seconds. In such a case, there are no seconds passing in the movie universe. The fastest observable change is a minute. Anything faster cannot happen or be observed in the movie universe.

1.3- Conclusion and The Next Level of the Story

A universe that
- contains space and time only experienced from inside.
- contains the past, present and the future simultaneously,
- exists like discrete time system at a very high frequency,

was the fundamental of my universe model when I graduated from the high school, in the late 80's. These were just ideas, I synthesized from what I read back then, without any knowledge of related mathematics.

After high school, I have studied Electronics and Communications Engineering at Istanbul Technical University. This is where I have taken lectures about electromagnetism, relativity, and quantum mechanics. I solved numerical problems about Schrödinger's equation or Heisenberg uncertainty. I saw all their proofs. That was the easy part of the story.

The difficult part was about the interpretation of the Quantum Mechanics. I could not understand why we had to explain everything with a probabilistic approach. In real life, we use probabilistic calculations for the cases, which we cannot model with all the detail but are deterministic.

Let's take the example of a dice. If you play dice with six faces, there is 1/6 chance to get one of the numbers between one and six. We all know that. You never know the result of a specific roll when the dice flies out of your hand. However, the whole process of playing dice is deterministic. If you could know the exact speed and angular momentum of the dice, its exact initial position, the fraction in the air and on the surface, you throw it to, the mechanics when it hits the surface, you could precisely predict the outcome. It is just so complex to make all the measurement and calculations that we don't dare to make it. A simple probabilistic approach is good enough to predict the result.

Anything we analyze by using probability is like this. The events such as the frequency of earthquakes on a specific fault, the average waiting duration at the cue in a supermarket etc. are causal and could be analyzed and predicted precisely if we had all the necessary data and the model but we don't.

As you may see, it was not clear to me, whether it is the same situation for quantum mechanics, too. Was everything totally deterministic but we could only make a probabilistic model because we could not go into the detail? Or was the universe random in quantum level?

Later, when I kept researching, I found out that it was not my fault not to understand this. This was the main discussion among the famous physicist back then. Some really thought that the nature of the universe is probabilistic, and some refused it. That's why Einstein said: "God does not play dice".

There are two components of the discussion which lead us to the fundamentals of the "Nature of Existence". Take an electron as example to explain them:

1. The probabilistic behavior of the electron: The electron doesn't fly around the nucleus of an atom, like the Moon orbiting the Earth. It just shows up randomly around the nucleus. It is truly random and can be modeled with a probability function based on its wave function.

2. The particle-wave duality: The electron behaves sometimes like a wave (energy) and sometimes like a particle (an object with mass). This also looks kind of random, but we can manipulate it under some conditions and force it behave like particle or wave. So, it is either particle or wave based on the circumstances. It appears as one of them randomly in its nature.

Quantum mechanics has surprising outcomes like these two which look unreasonable or look like paradoxes or dilemmas but there are real devices, working based on Quantum Mechanics equations. So, they must be valid. Maybe our perception about them is wrong. There must be another way of explaining them, a new way to look at the universe and a new story to tell about these equations.

In April 2020, when the world had to stay at Home because of Covid-19 Pandemic, I had the chance to read a new book about the Quantum Mechanics. Simultaneously, I was reading

Physics lecture notes[5] of Prof. Mark Rzchowski at University of Wisconsin-Madison. There was a slide about the electrons' random movement around the nucleus containing some notes and a graph of a probability distribution function on a coordinate system. On such graphs[6], the horizontal x-axis always represents the location of the electron, the space itself. So, the probabilistic function and the space are drawn as a whole, touching each other in such graphs.

Suddenly, an idea occurred in my mind. What if they don't touch each other? What if the wave of the electron is not dependent on the Space, so that it can randomly interact with it? What if the wave of the electron is NOT in space, NOT in the universe?

Let's use dice analogy to understand these questions. If you play crabs, you play dice again and again. Assume a crabs table is the universe and the dice represents the electron's probabilistic behavior. Whenever you play dice, you take them from the crabs table, shake in your hand, throw to the table again and have a result. The two dice are not on the table (not in the universe) during the process but whenever they get in touch with the table, they create a result on it.

What if it is the same for the electrons and their waves? Maybe the wave exists in a totally different medium, place whatever and it somehow interacts with the universe or space-time and creates a result? I realized that this could explain the strange things about the quantum mechanics

---

[5] The notes were published on the website of the university (https://uw.physics.wisc.edu/~rzchowski/phy107/lectureNotes.htm), but they no longer available.
[6] You can find a similar graph in Figure-9, Section 3.3

which conflict with our common sense like the wave/particle duality or double slit experiment cases. That's how the idea of the "Double Existence Model (DEM)" was born.

After that, I started a research and ideation process to check the compatibility of the model with known physical theories and tried to evolve it further.

DEM doesn't contain or claim a new physical theory. On the contrary, it uses the "accepted theories and information", and reverse-engineer them to establish a model about the structure of the universe, based on the main idea explained above. It includes definitions and explanations about space-time, quantum level events, Big Bang, time, and time dilation which lead to the description of our existence and its reality.

My personal goal has always been to comprehend the universe as a whole structure and understand the nature of our existence. The Double Existence Model certainly involves many answers about this. So, on one hand, you can position this book as a philosophical interpretation of physics laws to understand our reality. On the other hand, if this model has the potential to be used scientifically, it may help us to develop new theories someday, in the future.

One last word. The scope of the book could be much bigger. There are two main chapters about macro universe (relativity, Big Bang, time etc.) and micro universe (quantum mechanics). These were two useful topics to understand the model and make deductions about our reality of it. I could go into more detail and add more chapters, but I didn't because of two main reasons. First, they wouldn't serve the target of the book. Second, the book would be thousands of pages and become unreadable.

Thus, there are things which can be added to the book to extend its scope and define the model in more detail. On the other hand, of course, any idea in this book is subject to improve, correct, modify, evolve... This is the first version of the model which should serve to make us start to think about itself.

2- Where Do We Stand?

2.1- Scientific Methodology

The story of physics has been at its most exciting stage for the last 150 years. Before then, it was mostly based on mechanical systems which are directly part of our daily experience or within our limits of observation. We were dealing with things we could see with a naked eye or the worst case, by telescopes. It was about objects around us, planets, stars, and their movement. Careful people with high intelligence made observations and developed important theories.

When it came to the age of electricity, electromagnetic fields, subatomic forces, quantum mechanics, our world changed much. In this new phase, we must understand things, we cannot see and sometimes we cannot even imagine. Some of them are not part of our daily experience and some of them do conflict with our common sense. Furthermore, we deal with much more difficult questions because we almost came to the edge of the universe. Next things, scientists are working on are beyond our 3D geometric perception limits, smaller than anything we can measure and beyond our daily experience and understanding.

On the other hand, science begins with observation. Figure-1 shows the flow diagram of scientific methodology. Let's use the gravity case as an example. We observe the famous apple falling to the ground and ask the question: "Why do things fall down?". You can add other questions to go further. Why does the apple fall always down, not any other direction? What is the rule about the velocity vs. height? What is the rule about duration of the fall and the height?

The third, fourth and fifth boxes in the graph are about creating a theory which answers these questions. The hypothesis is a logical explanation about why things fall. This is the ideation step. You observe, question, and then find a logical answer. Then, at step 4, you formulate a testable prediction. In physics, this is generally a mathematical model you formulate based on your hypothesis. It's the key in the methodology. Any scientific hypothesis must make testable predictions so that its validity can be tested and disapproved or verified. At step 5, you run your experiments and see whether the mathematical model checks out. If everything is fine, you have a proven theory; if not, you may have to go back and modify your hypothesis or even find a new one. This is what science is about. Make observations, create ideas, and verify by performing experiments, objectively. It is the only logical way of doing science.

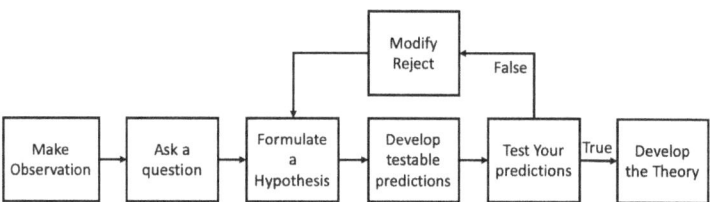

Figure-1. The flow of Scientific Methodology. The story begins with an observation and ends with a theory if it is a successful case. The observation leads to a hypothesis, and it is the base of a testable mathematical model. If it is verified, it becomes an accepted theory.

However, this doesn't always give us the "ultimate" correct answer. Sometimes, science gives us the best answer we can achieve which represents the truth to some level. It is never totally wrong but can be partially true or may only approximate the truth under specific conditions. Figure-1 shows the flow of development of one single theory. The full circle of scientific methodology includes one more feedback arrow from the theory box to the observation box. The theory itself, even if verified with experiments, can go back to the observation stage. This would trigger the whole process again which could result modification of the theory or could even lead to a new one. One of the main reasons why this can happen is that the mathematical model based on the hypothesis can work in the tests and even in our daily life, but it may not still represent the universal truth which is applicable in all possible conditions. So, to obtain a correct prediction model i.e., working mathematics is a must to have a valid theory but it is not a guarantee.

Let's analyze this using a sample case about gravity. I'll use a three-level story and therefore add an imaginary scientist with an imaginary theory about gravity in addition to Newton and Einstein. Let's call our imaginary scientist Karl Weber. He was born in 1224AD and developed his own theory in 1264, before Newton did. Weber is a clever, analytic person but due the level of knowledge and practical conditions of his time he was not able to develop a good model as Newton did. He could make simpler observations; he didn't think as universal as Newton did and built a model which was focused just on the gravity on Earth. He made observations and initial experiments to build his hypothesis and a mathematical model. Figure-2 shows his work according to the scientific methodology[7]. To focus on the details, only the first four steps are shown.

Weber didn't have any clue that gravity could be something universal. His observations and experience showed him that everything was only falling downwards to the Earth, not to any other direction. So, he thought there was something special about Earth. He might even think that it was in the center of the universe. In this case, it was so logical that Earth had a specific power of attraction. Everything was located around our world. It held them together with its power. The things which were very close, were affected even more and they fell. His hypothesis was based on the idea: "The Earth attracts everything". Hypothetically, let's assume he could distinguish mass and weight. In this case, he would say, the weight was always ten times the mass. Very interestingly, he would also find out the coefficient between velocity and fall duration was ten, too. What a magical number!

Anyone who has an idea about Newtonian mechanics and gravity model knows that Weber's model represents a special case of Newton's model which is only true on Earth. According to Newton, all objects with mass attract each other. This power is not a magical gift of Earth. The strength of the gravitational power is proportional with the mass of the object or planet. So, the static coefficient ten in Weber's model is specific to Earth. In fact, it is a variable which has different values on other planets.

---

[7] g, the gravitational acceleration of Earth will be assumed as $10m/s^2$ in the example.

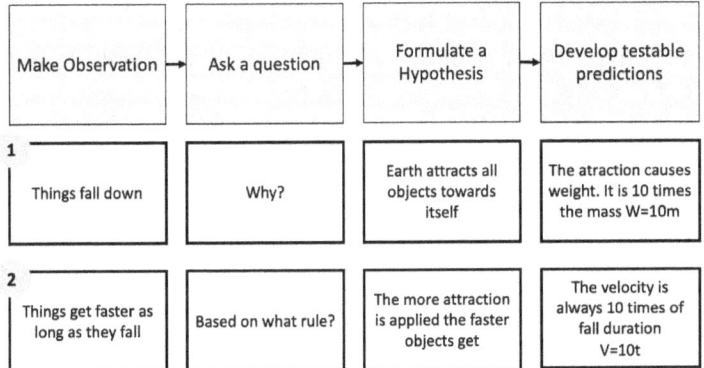

Figure-2. Weber's model of gravity. He observed things falling and decided, that was something about Earth attracting everything towards itself. The attraction causes weight, and the duration of the fall determines the velocity of the objects. His initial measurements showed him the coefficient between mass and weight is equal the coefficient between fall velocity and duration. It's ten.

On the other hand, if we lived in the age of Weber, we would make successful tests proving his model and would accept his theory, gravity is a specific power of Earth. This would be our vision about the universe. We would build mechanical devices like cannons, and they would work perfectly because everything would be working on Earth's surface. We would be quite confident about our theory if our ecosystem stayed on our world. However, if we tried to build a spaceship and calculate its route, we'd fail because we would be neglecting the gravity effects of the Sun, the Moon, and other planets in the Solar System. We'd need Newton's model to make the correct calculations. At that point, our vision about the universe would change. We'd find out that the gravity was not specific to Earth. All the planets, stars and galaxies attract each other in an equal game.

When Einstein comes to the picture, we see that Newton's main hypothesis is fundamentally wrong. Relativity tells us that there is no force of attraction between objects with mass. Objects with mass bend the space around them, so that other objects around them move towards these objects. Let's give an example. The Moon is not attracted by Earth. It moves in a straight line on its own way. That straight line is bent by Earth so that the Moon draws a circle around it, in other words it keeps orbiting around our world. This is a totally different universe compared to Newton's and changes its story.

On the mathematical model part of the story, relativity tells us, if things move fast (at a comparable speed to the speed of light) their movement cannot be modeled with Newton's rules, but it works when objects move slow. Mathematically, Newtonian mechanics is just an approximation of a more general model (Relativity).

The point is all the three theories would give the same mathematical result if we used them to calculate the speed of a falling apple from the tree, but they tell us different stories about how the universe works. In Figure-3, you can see how the three models are compared with this perspective.

As you can see, science is kind of journey along these four steps. If you can improve your observation skills and tools, you may have the chance to ask more clever or detailed questions and find improved or new hypotheses which lead to a more accurate and generalized mathematical model.

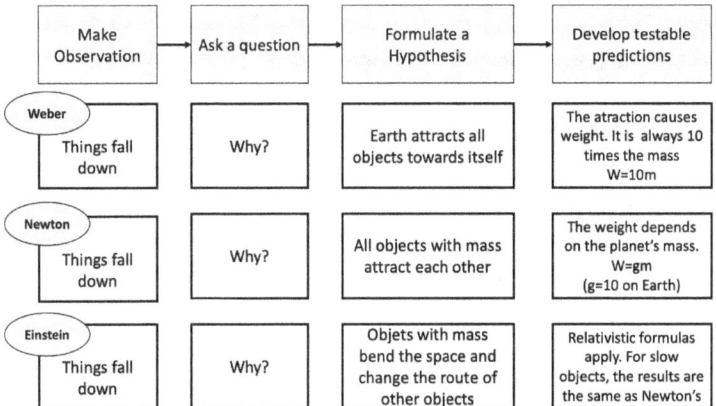

Figure-3. Comparison of three theories. All the three theories give the same result for the falling apple on Earth. The important thing is, their stories about the truth, the hypothesis they claim, is totally different. They describe different types of universes.

Mathematics is the language of the universe i.e., of science. Thus, the concrete scientific story is written in the 4th column of Figure-3, but let's focus the 3rd column, the hypothesis column. This is the most important step because it's the link between the observation and the theory. It's the lead which takes us to the 4th column. Mathematical model is the output of the hypothesis. More importantly, it's the main idea in plain English. So, it's what we think and know about the universe.

For example, in Weber's case, the hypothesis tells us that the Earth is in the center of the universe, and it has a specific attraction power which holds all the universe in order. In Einstein's universe, the space is like a big amount of invisible foam which is dynamically shaped by the objects in it all the time. There are almost no straight lines in Einstein's universe, every line is curved by some objects. These are two totally different visions about the universe and its existence.

The content in the hypothesis column is what we are curious about and what we want to know to understand about our existence. It's the core story of the universe and of us. Mathematics is only used to confirm that story. So, if you want to understand what physics or science is telling us, you should read the third column.

2.2- The Search for "The Theory of Everything"

If you are interested in physics, you know why it is important to find this unified theory about the four forces in the universe. This is the ultimate theory which should explain the fundamentals of the universe. Everything about universe should be defined by a mathematical model with such a theory so that we can make prediction about anything. This prediction refers to the 4th column of the scientific process and should describe the whole history and future of the universe.

Let's cover briefly what is it about exactly and why it is so important. There are four fundamental forces in the universe. Everybody is familiar with two of them.

The first one is the gravity. We experience it during our daily lives and know how it affects us. The way it works in universe, is an interesting story. At first, it was thought that it was an attraction force, but Einstein showed that it is geometric distortion of space caused by the objects with mass.

The second one is the electromagnetic force. It's the effect which causes electricity and all the related phenomena we know. Magnets, compasses, electro motors, radio & TV broadcasts, mobile phones etc.

The 3rd and 4th ones are called the Weak and Strong forces. These two have only effect within an atom and they hold the protons and neutrons together in the nucleus so that they don't fly away. They are not a part of our daily experience, but they enable our existence. Without them, the atom nucleus wouldn't exist, which means that the atoms wouldn't exist.

The theory of everything is a concept which would unify these four forces in one theory by using a common mathematical model and in a common story (Hypothesis). These forces determine all the movements and change in the universe. If we could understand their background, define them in the most general form, understand the relation between them, this would help us to understand the universe. We would exactly know how the universe is formed and how it works. This could lead to understand our existence. That's why it is so important to find "the theory of everything".

Physics has done a very successful job about defining the macro world (with Relativity) and the micro world (with Quantum Mechanics, QED, QFT, the Standard Model) separately until today. The problem is that the stories of them are very different. So, it is a big challenge to bring the macro and the micro cosmos (or the Relativity and the Quantum worlds) together and draw one big picture for these two.

The main hypotheses and the stories of these two theories are different. The relativity describes a fully deterministic universe where quantum tells us, all the mechanics inside an atom is about probabilistic functions. As a result of it, their mathematical formulations are not compatible. You cannot predict the movement of an electron by applying relativity model.

On the other hand, all the objects and events occur in the same universe, in the same environment. So, logically, there must be a common ground where these two theories are aligned.

Of course, there are valuable scientific studies in this area with important outcomes. The most popular one is the String

Theory. It is a very good example to understand the difficulty of defining a unified theory and to see how we need to go further. Thus, I'll very briefly describe what String Theory tells us about the universe and how it came to that point.

In summary, the main hypothesis of String Theory is that every kind of energy in the universe is caused by the vibration of very small strings. There are different types of strings which vibrate differently and produce different types of energy fields. These energy fields are the source of everything we observe as particles or energy waves in the universe.

The second important idea is to explain all four fundamental forces as geometric effects. As mentioned, Einstein has already showed that gravity was not an attraction force, it was a geometric distortion of space-time. Scientists working on string theory assumed that the other three forces could be similar. So, they thought that the other three forces could be caused by geometric distortions of some additional dimensions (addition to our 3D space + 1D time dimensions). The idea was that these additional dimensions were so small that we couldn't experience them. Many different models with different number of additional dimensions were tried between 5D and 11D universe models. They all are eliminated in time and the reason was, they didn't have consistent finite mathematical results. The 11D model is the first one which lead to finite and consistent values which means logical results. So, we can accept that an 11D universe model defined by String Theory could be a way to analyze the universe, but we may never be sure that it is the ultimate and solid truth because we will never be able to run an experiment to see whether there are really eleven dimensions.

Indeed, it is also possible that there are other alternatives which could give consistent mathematical results. For example, we could expand the number of dimensions to 12 and obtain logical outcomes. If you think that way, why wouldn't it be possible to have 13, 23 or 100 dimensions? Or there could be totally different models than String Theory which could also propose consistent mathematical outcomes, such as Loop Quantum Gravity.

The point here is to understand the level of difficulty of working on these topics. We came to such a point that we made a new shortcut in scientific methodology because we had to. Normally, based on observations, we must develop a hypothesis which should lead to a mathematical model. e.g., the number of dimensions in the universe. However, what we did in this case was to use mathematical consistency to eliminate the wrong assumptions and obtain some "possible" and logical results. So, hypothesis is modified by mathematical consistency, not by experimental outcomes. As like as not we should say that the hypothesis is "defined" by mathematics because we have no preliminary idea about the number of dimensions in the universe at the first place.

You can see this flow in Figure-4. The reverse arrow from testable model to hypothesis box represents the feedback, modifying the hypothesis based on mathematical consistency. While researching the limits of the universe, this is a tool we need to depend on which is useful and maybe the only thing we have in some cases.

The number of dimensions of the universe is just an example. There are so many other things we cannot experiment, observe, or perceive directly.

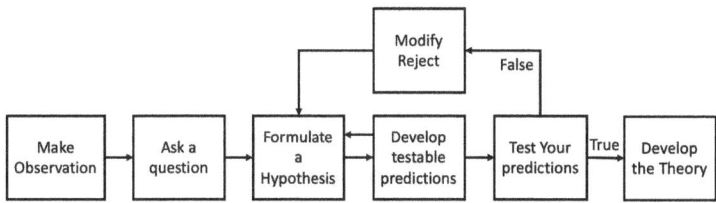

Figure-4. The Shortcut. In the search of the unified theory, we don't have the ability to test everything. So, in some cases, the mathematical consistency helps us to eliminate and filter the alternatives. The hypothesis is modified not by experimental results but by the prediction model itself. Sometimes this mathematical model is the only source to form a hypothesis because, as in the given example, there is no possible input to the observation column about the additional number of dimensions in the universe.

The theory of everything is the most important puzzle in Physics. To establish such a theory, we need solid hypotheses. Unfortunately, we cannot make end-to-end observations and create new hypotheses on top of them. Some parts of it are beyond our limits. That has been one of our biggest obstacles up to now.

2.3- Where does DEM stand?

As a curious person about the universe and trying to understand how and why it exists the way it does, I'm concentrated on its core story which is the hypothesis level. What does the universe tell me? How does it work? The mathematics is useful to prove it and serves engineering purposes, but I'm more excited about what it tells me.

At the end of the day, that's what we need to know to understand what's going on. What is time, why objects can bend the space-time, how is it possible that the macro cosmos is causal, but the micro cosmos is probabilistic?

I can add many questions like these. The answers to such questions constitute the story of the universe. They can be found by testing and proving hypotheses and completing the theories. As explained in the previous section, testing, and proving part is not always possible in this level of science and sometimes, the stories can only be completed with mathematical studies, not with observations.

However, if we had a full end to end story about the universe, a set of hypotheses which comply with each other and with what we know and accept as scientific truth, this could help us understand the universe better and lead to new or improved mathematical models. A unified theory, the theory of everything and its mathematical model should depend on a unified set of hypotheses which describe the universe in one story, in one model at micro and macro levels. Instead of trying to unify two very different stories at mathematical level, maybe we should write one common story in the background and start there.

Looking from the right way to the universe, synthesizing the information we already know with a new perspective could help us to write a unified story about the universe, The Story of Everything.

This is what the Double Existence Model is. The book is about the analysis of a universal model. The model is constituted by some fundamental hypotheses. It's about the structure of the universe.

Figure-5 shows where the model stands according to the scientific methodology. The model is based on known facts and observations. It asks why/how questions and proposes answers. The scope of this book and the model itself is concentrated on establishing a unified story (the set of hypotheses) about the universal structure and the reality of our existence.

The third column of Figure-5 is about the story of the universe. It tells us how it works and why is everything like this. Chapter 3 explains the fundamentals of the model. This is where we will see how the universe is structured according to the Double Existence model and its main hypotheses. Chapter 4 and 5 are analyzing the Macro and Micro worlds respectively, comparing them to the model. You will get answers to the questions like "what is the time, why it slows down if we move fast? How do objects bend the space time? What is the root cause of expanding universe?" And many others.

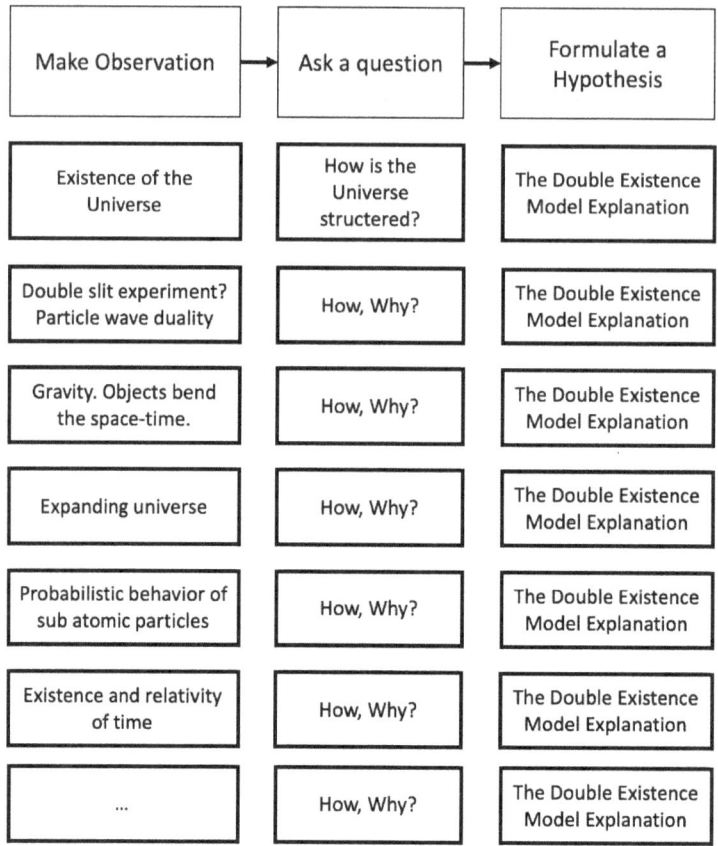

Figure-5. The Double Existence Model in the methodology process. The first line represents the main hypothesis about the structure of the universe which also explains the properties of the universe and completes a set of hypotheses. The core of this book is the story of the universe, defined by the 3rd column of this table.

3- The Double Existence Model

3.1 Introduction

Einstein's work showed that matter and energy can be converted to each other. So, in fact, they are just another form of each other. Everything "in the universe" is either energy or matter. The Standard Model suggests that there are different types of energy fields "in the universe" which cause the presence of particles, like the electron fields. This means everything is a kind of energy, even matter is a type of presentation of energy fields. This is something very difficult to comprehend if you are not an expert in physics. On the other hand, all the answers to the questions about the universe and its origin lie here. It's about the dance of energy, matter, and space time.

The Double Existence Model (DEM) is a set of hypotheses which proposes an explanation of the relationship between energy, matter, and space time. This leads us to many answers about the structure and the mechanism of the universe.

DEM is a kind of reverse engineering of the accepted physics theories, and it uses them to define a common background story, The Story of Everything, using a new perspective. This helps us to answer some questions which were not clearly answered before.

- What is time? What is its origin?
- What is the origin of speed of light?
- What caused Big Bang?
- What is the relation between the causal macro world and probabilistic micro world?

- Why is everything quantized?
- What causes the relativistic phenomena like time dilation?
- What causes gravity?
- Are there parallel universes?
- What is the reality of the present time we experience?
- Is "time travel" possible?

You will find concrete answers to all these questions in this book. The story defined by DEM has proposals for all of them. All the hypotheses and answers delivered by DEM depend on the background story. As mentioned before, this story is derived from the current physics theories by combining and interpreting them using the main assumption of DEM.

3.2- The Definition of Ultimate Universal Set

The Double Existence Model is an end-to-end set of hypotheses which describe the structure and mechanics of the universe as a whole. The universe is analyzed not just within itself but also referring to "beyond it". Thus, a new definition is required to describe the environment and the things which are not in the universe, including the universe itself, the Ultimate Universal Set U^U.

Hypothetically, there are many different possible and logical scenarios about what's beyond the universe. The universe could be a 4D Space-time balloon in a bigger 4D universe, or a subset in 8D Hyper-universe or one of one billion other parallel Universes etc. A true and scientific representation of such a "whole existence" should be independent of such scenarios. So, the definition of the U^U should avoid any kind of geometric, time or material-based references and must only describe the fundamental true state of existence.

This requirement leads us to the set representation. Sets only contain the list of their elements and don't define any relationship between them which makes the sets suitable to represent or describe the "whole existence".

The etymological origin of the word universe relates to concepts about "everything, all, all things". That's why this word is used to name the universe. On the other hand, all the existence including the universe itself and beyond should have an additional distinguishing descriptor where we'll use the word "Ultimate".

Combining all these inputs, The Ultimate Universal Set of "whole existence", U^U, is a set which has "everything that exists" as an element. The only element of U^U, as far as we know, is our universe. We have no information, whether any other elements exist in U^U. So, the mathematical description of U^U would be as of now,

$$U^U = \{The\ Universe\} \tag{1}$$

or in Venn diagram, as in Figure-6.

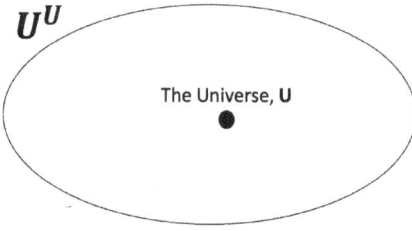

Figure-6. The Ultimate Universal set U^U, as far as we know.

3.3- Definition of The Double Existence Model

The Double Existence Model's main Hypothesis:

The matter/energy we observe and we are made of is not in the universe i.e., not in space-time.

The matter/energy is just a momentarily "projection" of an "external" Complex Energy Source (CES) on Space-time.

(H1)

Maybe the most generic postulate of science is that the universe includes everything in it. Everything is within the big volume that we call "Space". With Einstein's work, we realized that time was also a dimension and called it space-time. Space-time is the base of the universe and there is a total amount of "Matter + Energy" in it which may change via quantum level energy fluxes. In other words, Space-time is a container, there are "Energy & Matter" in it and we call all of these "the universe."

On the other hand, the Double Existence model suggests that the universe is constituted of two independent (but interacting) components:
- 4D Space-time (ST)
- The Complex Energy Source (CES)

43

Space-time and CES are two independent objects in U^U as shown in Figure-7.

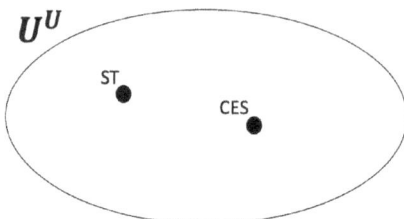

Figure-7. The Space-time and the Complex Energy Source in U^U.

The benefit of using the set description of U^U helps us here to present the structure with two components without defining any geometric or other physical relation between ST and CES. At this point of the book, the only focus is that ST and CES exist independently, and they do have a kind of interaction. No definition is made about how far or close they are or whether they are "in a kind of physical contact" or not.

The main roles and physical properties of these two components can be summarized as follows:

- Space Time is a kind of cosmic medium where the complex energy in CES can be projected under some conditions on it and get into interaction. In this case, ST reflects a kind of 4D shadow of complex energy either in the form of Energy or Matter, depending on the situation.

- CES is the source of matter and energy in the universe. It contains a big amount of "Complex Energy" which is constituted of various energy components with complex wave functions having more attributes and variables than we observe in ST. (e.g., having more dimensions, more complex components, functional components). Whenever a complex energy wave is projected on ST, we observe it as a matter or energy with measurable attributes. That's only a part of total energy in CES. Thus, there are more energy waves in CES than we observe in ST.

Based on these descriptions, for any given moment t_1, an amount of total energy E_1 in CES is projected on ST, and at the next moment t_2,

- some of E_1 will be projected again and continue its existence in ST.
- some of E_1 won't be projected again and continue to exist only in CES.
- some other energy waves of CES will be projected and appear in ST as particles or energy.

The second important hypothesis in the model is about the behavioral situation of the waves depending on their state of projection.

There are two states and two main rules:

If a complex energy wave in CES is projected on ST, it obeys the mechanical rules of ST. It acts either as a particle or as a wave and is observable in ST.

(H2)

If a complex energy wave is not projected on ST, it continues its existence in CES only, behaves as a complex wave according to CES mechanics and its wave function. It's not observable in ST.

(H2) defines the core rule set about how the universe works and is responsible of big part of the fundamental physical phenomena like, random structure of quantum mechanics, strange outcomes of experiments like the double slit experiment, the debate about the randomness vs. causality of the universe.

Figure-8 shows the way the projection occurs and how it affects the energy waves in CES, which explains the scope of (H2) in detail. Figure-8a presents the basic projection. The 3D red to yellow coloring represents the complex energy waves in a sample 3D CES environment. A part of the complex waves is projected on 2D-ST and creates a shadow in it. Figure-8b explains the feedback from ST to CES. Because the waves within the black square are projected on ST, they are also a part of it and they are affected by the mechanical rules in ST. Based on their projection mode, they can be displaced by any

46

force or reshaped into a specific 3D geometric form. All these effects will be reflected to the group of waves within the black square in CES environment which originally created that shadow. On the other hand, the waves outside the black square have no connection with ST and they act freely in CES environment according to their complex wave functions. Figure-8c is an example of a solid object (a cylinder) and its source waves in CES. Because of the correlation between the source waves in CES and the shadow in ST, the complex waves in CES become a correlated form when the object is shaped as a cylinder. The form of the waves in CES is not necessarily a cylinder at the same size but it's something correlated which has a cylindrical shadow in ST. Thus, this is a symbolic representation.

The interaction between ST and CES, as defined above, is such a mechanism that ST encloses, captures, imprisons the complex waves at the moments they are projected on ST. The mechanical rules of Space-time are applied on them. Whenever the waves are released by ST, they act in CES as pure complex waves.

ST and CES are defined as two independently existing objects and, no geometric or physical relation between ST and CES is defined. To use an analogy for the sake of expressional simplicity, it can be stated that they sometimes "touch" each other at some points and interact. Thus, at some point they have a kind of physical contact in the environment of U^U, whatever that can be.

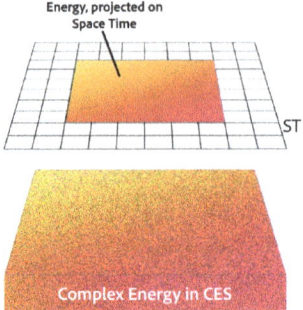

Figure-8a. Basic presentation of sample projection of Complex energy on ST which is shown as a 2D surface. A part of the complex energy, floating in CES, is projected on ST and created an energy field which is observable in ST.

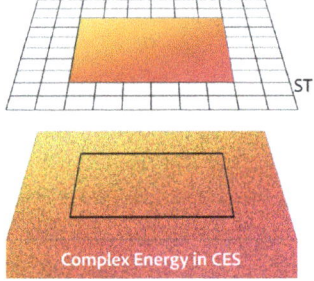

Figure-8b. The part of CES which is shown in black square is projected on ST. The complex waves are in CES but have a "shadow" in ST. Any "physical effect" of ST on the shadow is reflected to the related complex energy waves within the black square.

48

Figure-8c. Another example with a solid object. If the projection of a specific group of complex energy waves is a cylinder-shaped solid object in ST environment, then the group of waves in CES has "correlated" shape in its environment (not necessarily the same shape) and if the cylinder is moved in ST, the group of waves moves in CES.

The point is the hypothesis (H1) "CES and the complex energy waves not being in ST" doesn't forbid any kind of physical interaction but it suggests that the waves have an independent environment and rule set in CES on their own and they have more definitive attributes than we observe in ST.

One of the most important outcomes of (H1) and (H2) is related to the most famous physics equation, $E = mc^2$. This equation tells us that energy can be transformed into matter and the matter can be transformed into energy.

Thus, energy is defined as a form of matter and matter as a form of energy. This expression is correct but according to the

Double Existence Model, a generalized version represents the truth more accurately. Matter and energy are not just other forms of each other. Instead, as stated in (H3):

Matter and energy are two projection forms of Complex Energy (their origin) on Space-time. (H3)

(H3) explains the convertibility of matter and energy by their origin in a simple form. Originally, both are all complex energy waves in CES. Their modes, forms of projection on ST can be different or varying in time but they all are the same in the first place, complex energy waves with their complex wave functions in CES.

Here is a sample case to explain the relation of CES and ST. Consider an electron with a probability distribution function of its location along the x-axis as shown in Figure-9. According to the distribution function, the electron can be at point E at a given time point t_1 and then appear at point F at t_2. This is how it really works in quantum world. The weird thing is, as well known, the electron can never be at points A, B and C, but somehow it can pass through the space and appear at F. If the x-axis was a street and the electron was a man walking from left to right, he could walk from point E to F, but he had to pass the points A, B and C.

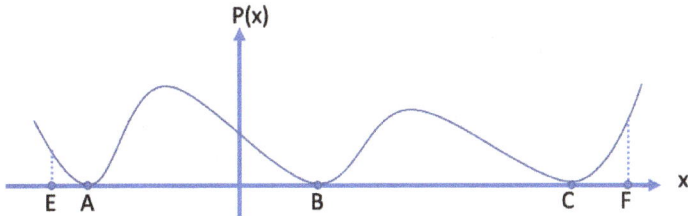

Figure-9. The probability distribution function of a sample electron. The electron can never be at points A, B and C because the probabilities of being at those points are zero. However, it can travel from point E to F.

This is one of the most known cases about Quantum Mechanics which conflicts with our daily experience and common sense. This conflict leads to many questions. Why does the electron appear on random points? Why does it sometimes act as particle or sometimes as wave? What causes either of them, if the wave is always a part of Space-time and nothing changes? How can it skip points A, B and C? What kind of reality do we obtain when the wave function collapses?

These kind of questions or dilemmas are mostly caused by looking at the subject with a mixed probability/causality window without the necessary assumption set. Why should the wave function collapse if it is applied at moment of time? Or why should the universe be divided into two realities when it collapses, creating parallel universes for either solution of the wave function (like one universe where Schrödinger's cat is dead, and in another one it is alive)?

The Dual Existence Model proposes another perspective in (H1), (H2) and (H3). The complex wave of the electron

51

continues to exist in CES forever and it moves according to its wave function in CES environment. Its presence in ST is just a momentarily projection, a shadow. That shadow is one sample solution of its wave function which will continue to exist in CES afterwards and will be projected again later.

The journey of the electron's presence according to DEM is as follows (Figure-10):

1. The electron's complex wave is in CES. The wave has its own Complex Wave Function $f_{CE}(\vec{x})$[8] which is the origin of P(x). The wave continues its existence in CES and moves according to $f_{CE}(\vec{x})$.

2. At time point t_1, the wave is projected on ST, and it appears as an electron at point E where E is one possible solution of P(x) (and $f_{CE}(\vec{x})$). While the electron is at point E in ST, its wave function $f_{CE}(\vec{x})$ still exists and valid in CES, independently.

3. The electron disappears in ST, and it is only in CES. $f_{CE}(\vec{x})$ controls the wave.

4. At t_2, it appears again in ST, this time at point F, another solution of P(x). Because the electron was not in ST between time points t_1 and t_2, the mechanical logic of ST was not applicable. The P(x) in Figure-4 is the result of the projection of $f_{CE}(\vec{x})$ on ST, and it's in charge of determining the electron's new position independent of its previous position at t_1 and independent of the geometric relation of points E and F in ST.

---

[8] The argument $(\vec{x})$ is used just as a generic and symbolic vectoral expression. It shouldn't be taken as a 3D geometric expression related to Space-time.

Figure-10 shows this flow. The probability function P(x) is shown symbolically separated from the x-axis to emphasize, it's not in ST. It's shown like a hovering standalone wave (being in CES). X-axis (ST) is standing below it and there are two different time points: t_1 and t_2.

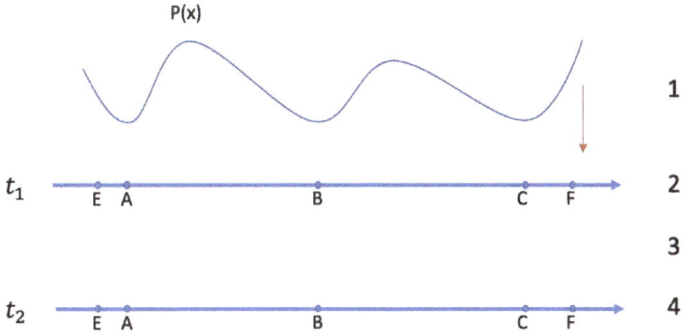

Figure-10. The displacement of the electron from point E to F. The electron's complex wave is in CES at Step-1. Symbolically, it's shown apart from the x-axis hovering in air. Assume, it moves down and hits the x-axis (ST) first time, at t_1. This is Step-2 and the electron appears at point E. Then the wave moves on and gets separated from the x-axis and disappears from ST in Step-3. At Step-4, the wave hits again the x-axis and this time, it appears at another point F which is suitable to its wave function. So, its appearances have no dependency on the geometric relation of E and F in ST environment. The projection process and source wave function are totally independent of ST and its geometry.

As shown in the sample, the complex wave is always present in CES and its wave function is effective. Whenever the wave is projected, the electron appears in ST as a particle and can be affected by ST mechanical rules while its wave is still in CES environment. The moment it disappears from ST, it continues its existence in CES as a wave without being affected by ST mechanics anymore. In other words, it goes somewhere in CES (while not being projected on ST) so that it appears at point F at the next moment t_2 of its appearance in ST.

This mechanism helps to understand and clarify many physical phenomena. Anything categorized as particle-wave dilemma can be explained with this model. The behavior of sub-atomic particles gets logical when they are assumed to be projections on ST which continue to exist as wave function in CES, simultaneously.

The summary of the Double Existence Model's definition and its main characteristic properties can be listed as follows:

1. The "physical universe" we live in, and we observe, is constituted of two independent components.
   a. Spacetime (ST)
   b. Complex Energy Source (CES)

2. The matter and energy in the universe are projections of Complex Energy in CES on Space Time.

3. CES is the independent source of energy and matter. It represents the fundamental existence. The complex wave functions in ST are defined by the complex wave functions in CES environment.

4. CES and ST interact and constitute the universe as we know, by
   a. Creating all physical rules, we observe in the universe (in ST)
   b. Creating interaction from the mechanics in ST to CES, so the complex waves in CES are affected by the mechanical rules in ST.

5. The existence of our universe is a duality, originated in CES and projected on ST.

3.4- Main Mechanics of the Universe

To establish the mechanical principals of the universe as a whole system, the hypotheses of the model can be used to define a proper format. Such a format should represent all the fundamental properties of standalone elements and the relation between them. The complex waves in CES environment, their way of projection on ST, the mechanics applied to them in ST and the reverse effect from ST to CES should be taken into consideration. The complete mechanical structure must also address the probabilistic behavior of sub-atomic world and merge it in a logical way with the causality of the macro world, which would also solve the problem of combining them in one scenario. If all we observe is true realities, indeed they are, then there must be a common background story and system to combine them.

Figure-11 shows the Universal Mechanical System, proposed by the Double Existence Model. It includes the CES, ST and complex waves. It fully describes the complete flow of energy, matter, the effective systems in place and most importantly where each system is in effect. This is very important because the defined relation between Quantum and Macro worlds help us to understand and solve the probability vs. causality dilemma.

According to (H1) the source of energy and matter is the CES. That's why the flow in Figure-11 starts with complex waves in CES. They exist in CES, and each wave has a complex wave function $f_{CE}(\vec{x})$.

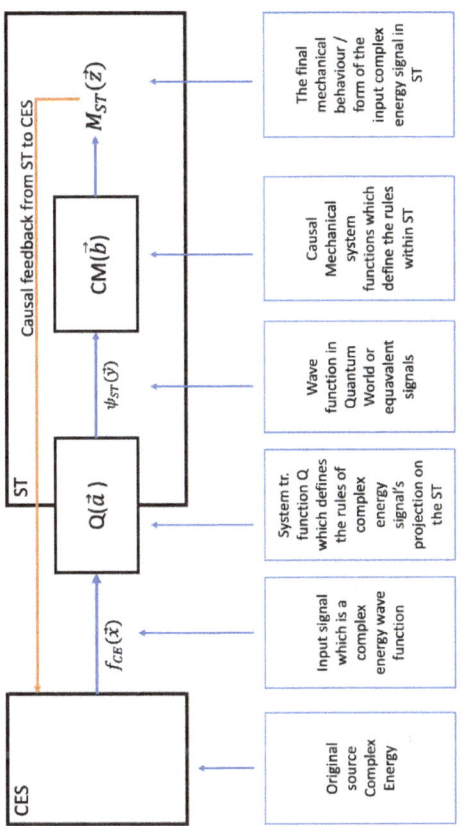

Figure-11. The Universal Mechanical System Chart. The complex waves, originated in CES, with complex wave function $f_{CE}(\vec{x})$ are projected on ST. The projection is run by a Quantum level rule set Q which transforms the wave function into another complex and probabilistic wave function $\psi_{ST}(\vec{y})$. The energy waves or particles with $\psi_{ST}(\vec{y})$ are parts of physical systems or objects in ST. Classical Mechanics rule set CM is applied to the systems they build, which is totally causal. These states (position, momentum, shape etc.) create feed back to the original complex waves in CES.

One important issue is that we can't know whether $f_{CE}(\vec{x})$ is a probabilistic function or not. Both scenarios are equally possible. CES is an environment beyond ST and may have many different and additional "physical" attributes like extra dimensions (or no dimensions), other mechanical rule sets etc. These "extras" beyond our observation and analysis capabilities may cause our perception of probabilistic nature for wave functions $\psi_{ST}(\vec{y})$ which is a projection of $f_{CE}(\vec{x})$ on ST.

In summary, the environment and the character of complex wave functions in CES might be causal but their projection on ST may look probabilistic to us. It is also possible that, $f_{CE}(\vec{x})$ is a probabilistic function by its nature in CES.

Whenever the complex waves are projected on ST, they go through a system "Q" with a set of rules which convert them so that they can have presence in ST. Q is the quantum level transformation system with a transfer function $Q(\vec{a})$. This system generates the particles, the energy fields of particles (as in the Standard Model) and all the energy and waves we observe in quantum level. Q is the entry point of complex energy into ST, which works only in quantum level and defines the world on this level.

Combining all these in (H4):

> *The quantum mechanics (Q) is about the projection of complex energy on ST, the rule set of its behavior while entering ST.*   (H4)

(H4) mainly suggests that the quantum world and the mechanics in that level reflect the projection mechanism of complex energy from CES on ST. The system Q in Figure-6 is a kind of gatekeeper. It transforms and accepts complex energy waves into ST, probably removes some of their attributes during projection.

One important point to mention is that the flow in Figure-6 doesn't show a signal flying away from its source, going through different systems and end up somewhere. This is the usual way how such flowcharts would be interpreted. However, in the model, the complex wave never leaves CES. On the contrary, it always stays in CES but its shadow is projected on ST going through the system in Figure-6.

The output of system Q has various elements (electrons, photons etc.) with their own complex wave functions $\psi_{ST}(\vec{y})$. They build systems, atoms, molecules and finally objects at macro level where the Causal Mechanical system (CM) is in charge. This is where our common sense, experience and causal logic start working. The output of this system is everything we can see around in our daily lives: Cars, trees, computers, books, apples, and ourselves.

Take the falling apple as example. While the quantum level system Q continues to accept projected complex waves in, which enables the apple to exist in ST, the CM system (because of gravity) makes it fall from the tree. The feedback from ST to CES carries the physical effect of the fall. At the next moment of the fall, the complex wave functions of the waves in CES which constitute the atoms of the apple, have a new set of initial or boundary conditions (keeping their own probabilistic character in micro level). Thanks to the new initial/boundary conditions of the wave functions, their solution is also modified. This means, the complex waves of the apple in CES fall from the tree, too.

> *The classical/casual mechanics CM is about the systems which are created by complex energy waves' projection on ST by the system Q.*   (H5)

> *Probabilistic Quantum World (Q) and causal Classical Mechanics (CM) have a kind of "in series" relation with respect to their scope.*   (H6)
>
> *CM defines the second step of the rule set (valid in macro level) for the complex waves which entered the ST via Q (Step 1).*

(H4), (H5) and (H6) are the second trilogy of the Double Existence model and define the main rule set of the Universal Mechanical System. According to the model, quantum level physics is directly related to the mechanics of the projection of complex waves on ST. The way an electron moves around the nucleus of atom, random flips between particle/wave states behaviors etc. is about the entrance phase of complex energy into ST. All randomness is caused by the nature of the complex wave with $f_{CE}(\vec{x})$ and its projection process Q on ST. At the same time, at macro level, the rules of CM apply (any movement in space, getting a form etc.). This may create causal feedback from ST to CES. It runs the macro world, as we know it.

3.5- The Complex Wave Function in CES

One last analysis must be made before discussing the relation of the Double Existence Model with macro and micro universes. The theoretical structure of the complex wave function $f_{CE}(\vec{x})$ in CES needs further attention to understand the source of the existence and its effects on ST.

At the start point, there is no solid information about the environment and the content of CES. That's the reason why a set presentation was selected to show it in U^U. No "physical" or "geometrical" definition was made not to limit the model with coincidentally defined attributes and make false deductions. As mentioned before, CES can be a multi-dimensional environment or something totally different with no dimensions. Likewise, the $f_{CE}(\vec{x})$ functions may contain some of the components of their projection function $\psi_{ST}(\vec{y})$, or not. According to the model, the only relation we can define is:

$$\psi_{ST}(\vec{y}) = Q(\vec{a}) * f_{CE}(\vec{x}) \qquad (2)$$

There is no information about $Q(\vec{a})$ and $f_{CE}(\vec{x})$ available. $f_{CE}(\vec{x})$ has information about the complex waves in CES but as an unknown $Q(\vec{a})$ has modified it. It's not possible to reverse engineer $f_{CE}(\vec{x})$, using $\psi_{ST}(\vec{y})$[9].

---

[9] From this point on, $\psi_{ST}(\vec{y})$ will be used in a form, containing sinusoidal components only, for the sake of simplicity to design examples and make graphical analysis.

Based on the transformation characteristics of $Q(\vec{a})$,

1. $\psi_{ST}(\vec{y})$ and $f_{CE}(\vec{x})$ may have similar or even identical functional components and arguments in common.

2. $\psi_{ST}(\vec{y})$ and $f_{CE}(\vec{x})$ may have completely different functional structures.

In Case-1, they have some shared functions and behaviors. This means, $\psi_{ST}(\vec{y})$ is a kind of subset of $f_{CE}(\vec{x})$ with some linear modifications. At the same time, $f_{CE}(\vec{x})$ has many additional functional components which define it in CES environment fully. In this case, the source of the probabilistic nature of $\psi_{ST}(\vec{y})$, we observe in ST, is a derivation of the probabilistic nature of the complex waves with $f_{CE}(\vec{x})$ in CES.

In Case-2, these two wave functions have no common ground. $Q(\vec{a})$ is a fully non-linear transformation. It's not possible to reverse engineer from $\psi_{ST}(\vec{y})$ and make assumptions for $f_{CE}(\vec{x})$. In this case, the environment in CES and the nature of $f_{CE}(\vec{x})$ may be causal or probabilistic. That cannot be determined. The probabilistic behavior in ST may then be just our perception about a very complex environment's reflection on ST or a transformation effect of $Q(\vec{a})$.

In the rest of the book, Case-1 will be used to give examples, make graphical analysis and analogies, for the sake of simplicity. This doesn't change the model's main assumptions or effect the results that the model suggests.

The generic structure of $f_{CE}(\vec{x})$, assuming Case-1 is valid and $\psi_{ST}(\vec{y})$ is just a function with real sinusoidal components, is shown in (3).

$$f_{CE}(\vec{x}) = \sum_{k=1}^{n} A_k \sin(f_{k1} + \cdots + f_{km} + c_{k1} + \cdots + c_{kl} +$$
$$g_{k1}(\vec{x}) + \cdots + g_{kp}(\vec{x})) + \sum h(\vec{x}) \tag{3}$$

(3) only shows a simplified version of $f_{CE}(\vec{x})$ to represent the logic behind it, without putting every possible variable not to create visual discomfort. Normally, $f_{CE}(\vec{x})$ is at least expected to include imaginary versions of all the terms in (3) with probably more than one imaginary dimension (i, j, k, ...). This real part in (3) will be used only to analyze the relation between $f_{CE}(\vec{x})$ and $\psi_{ST}(\vec{y})$ as a simplified sample case.

The expression in (3), based on Case-1, includes a generalized version of the wave function $\psi_{ST}(\vec{y})$ because it is assumed to be a linearly transformed sub-set of $f_{CE}(\vec{x})$. Assuming a basic structure (a function with real sinusoidal components only) as example for $\psi_{ST}(\vec{y})$, $f_{CE}(\vec{x})$ has a wider range of sinusoidal functions where,

- There are "n" different sinusoidal functions with
  - "m" different frequencies (probably because there are at least m dimensions or directions in CES)
  - "l" different phases or constants
  - "p" different additional factors which are defined by unknown functions.
- The wave function we observe as $\psi_{ST}(\vec{y})$ includes a transformed subset of these sinusoidal functions after $f_{CE}(\vec{x})$ is projected on ST passing through $Q(\vec{a})$.

Additionally, (3) includes the term "+ $\sum h(\vec{x})$" which symbolizes all possible additional functional components of

$f_{CE}(\vec{x})$ which relate to its nature in CES and are not projected on ST as a variable or effect. These functions have no projection or relation to ST environment and probably are beyond our perception and definitions. Of course, there is also a possibility that $\sum h(\vec{x}) = 0$.

In summary, (3) doesn't include the whole expression of $f_{CE}(\vec{x})$, but represents the logic of its relationship with ST and $\psi_{ST}(\vec{y})$.

Figure-12 shows a very simple example of $f_{CE}(\vec{x})$. It's just a 1D signal with four sinusoidal components which constitute five different pieces of signals. This is just a basic example to make some analogies and explain the main logic. In reality, $f_{CE}(\vec{x})$ is assumed to have much more arguments, dimensions -if any-, and other type of functional components in CES environment than just the three variables of the sinus function, amplitude, frequency and phase.

Let's assume, the complex wave function $f_{CE}(\vec{x})$ in Figure-12 is the source energy of an electron in CES, as an example. The following list includes some sample mechanical scenarios how the Universal Mechanics works when this energy signal is projected on ST. These are only sample scenarios. Other combinations are also possible.

Let's assume only the signals (1) and (2) can have a presence in ST when this signal is projected because they are the only two signals which can interact with ST directly.

In this case,

- The mass of the electron is determined by the total energy of (1) and (2) after passing $Q(\vec{a})$.

- Signal (4) is never projected on ST. It cannot be directly observed and it is not known in ST environment. However, its phase determines the spin of the electron. Its phase can be 0 or π which makes the spin +1/2 or -1/2 respectively.

- The order of the signals (2)-(3)-(2) determine the negative charge of the electron. The protons have the (3)-(2)-(3) sequence and they have positive charge.

- The signals (1), (2) and (3) pass $Q(\vec{a})$, and have an influence on the wave function $\psi_{ST}(\vec{y})$ in ST environment.

- Signal (1) is the main origin signal of the electron field in ST environment. Its amplitude is related to the energy level of the electron. When a photon is absorbed by the electron and its energy level is changed, the amplitude of (1) increases in CES.

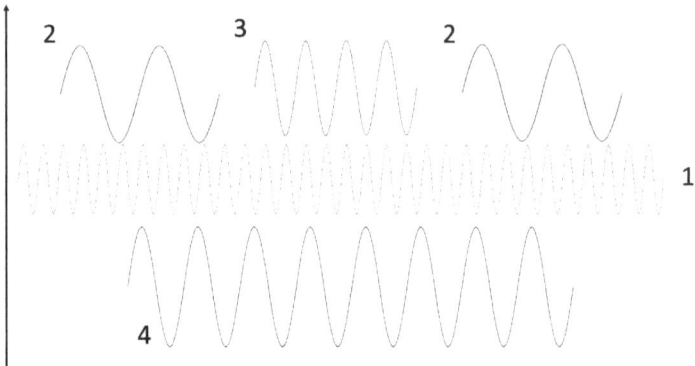

Figure-12. A 1D simple example of $f_{CE}(\vec{x})$ in CES. There are four different types of components in this structure which define different properties of the signal. Not all of them are necessarily projectable on ST. Some of them have direct effects in ST domain like determining the mass of a particle, some have indirect affects some have none.

We can also use the interpretation of $f_{CE}(\vec{x})$ according to Figure-12 to explain the mechanics in subatomic domain.

- Any behavior or interaction of subatomic particles can be analyzed by modelling the complex wave function as in Figure-12, as a sum of its components and analysis of these components' behaviors and interactions.

- As an example, any action which is shown by a Feynman diagram can be analyzed as;
  - Sharing
  - Exchanging
  - Addition
  - Neutralization
  - ...

of components (1), (2), (3), (4) with some other particle's $f_{CE}(\vec{x})$ components in CES domain. e.g., if an electron absorbs a photon, then its component 1 is added with the photon's component 1 and has a bigger amplitude.

3.6- Summary of the Double Existence Model

In summary, the description of the universe by the Double Existence Model is:

- There are two components which constitute the universe, the space-time and Complex Energy Source.

- No matter and energy are originally present in the universe. They are a projection of Complex Energy in CES on Space-time.

- When an energy wave is projected and has presence in ST,
  - it still exists as a complex energy in CES.
  - it is affected by the mechanical effects applied to it in ST.

- Matter and energy are two forms of projection of complex energy on space-time.

- The micro world where Quantum Mechanics is in charge is about Complex Energy being projected and taking presence as matter or energy in space-time. It's the rule set of entering the space-time.

- The macro world where classical mechanics is in charge is about the behavior of matter and energy which is projected on ST and is part of a system in space-time.

- The complex wave function $f_{CE}(\vec{x})$ includes many components in CES domain, some of them are projected on ST and cause observable results. The interaction in subatomic world is about the interaction of these wave components in CES domain.

4- Space-Time and Macro Cosmos

4.1- Big Bang and the creation of Space

Big Bang is the first moment of existence according to our understanding. It's defined as an explosion which started the existence and expansion of space. By rewinding the expanding universe back to that moment, the universe is assumed to have no volume, being a singularity or nothingness.

The universe model in Figure-7 shows the current structure which is valid after the Big Bang. What assumptions can be made about the moment of the Big Bang -or symbolically[10] just the moment before it- and how would it look like in U^U?

Logically, if the Big Bang is the first moment of space-time's existence, then ST can't exist "before" the Big Bang, at the symbolic moment of time t_{-1}. If anything related to ST existed at t_{-1}, then it must be something else.

This hypothetical subject is called space-time-Kernel (STK) in the model. STK is the origin or the root cause of space-time which became ST via Big Bang. On the other hand, the existence of CES is totally independent of ST's existence. Theoretically, it may or may not exist eternally. However, CES must exist at t_{-1}, so that it can take a role in Big Bang. It may be always there, or it may have started to exist just then. In both cases, it can be assumed that CES existed at t_{-1}.

---

[10] It's a symbolic phrase because time, as we know it, is not defined before Big Bang.

Combining all these assumptions and outputs, the U^U should look like in Figure-13 at t_{-1},

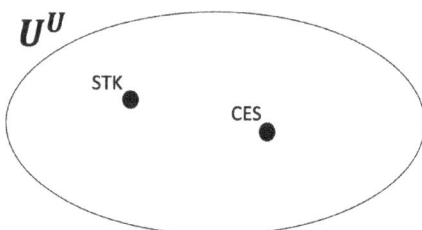

Figure-13. The Ultimate Universal Set "before" the Big Bang at t_{-1}. It includes CES and the Space-time Kernel, the origin of Space-time with no defined physical attributes.

Disclaimer: *"One important note about the CES and STK: The Double Existence Model focuses on the universe's general structure and tries to define its mechanics. The target is to understand the elements and mechanics of our existence. Chapter 4 focuses to explain the existence and properties of space and time. This analysis begins with Big Bang so that the creation and the structure of space and time can be defined. We cannot explain space-time related issues without defining how they are created with Big Bang. That's why a model is proposed to explain how Big Bang occurred as a starting point but it is never the goal of the model to explain why and how CES and STK exist or to define their origin."*

It can be stated that Figure-13 shows the "initial conditions" of Big Bang. The Complex Energy Source is an element of U^U as a standalone object and the Space-time Kernel is another element of U^U. There is no information about their status and the model doesn't refer to them.

The important thing is, if the universe is constituted of two components, ST and CES and if all the matter and energy in ST is a projection of the complex energy in CES on ST, then, using the metaphor "they touch/contact each other in a way",

> *Big Bang is the contact of the Complex Energy*   (H7)
> *Source with the Space-time Kernel.*

The contact of CES with STK, the Big Bang, created the space[11]. There is no evidence of what STK is. It can be a kind of "**object**" in U^U environment, or an "**event**" which interacted with CES, an independent "**energy wave**" which entered CES or anything. It can only be defined as one of the components of Big Bang. When STK and CES contacted, as Big Bang happened, the energy in CES interacted with STK and this created the Space. The main reason why space is formed and started to expand, is also the complex energy in CES.

Scientific studies and theories about Big Bang and the expansion of the universe suggest similar mechanisms. The first moments of the universe which was extremely hot, started to cool down and expand, the formation of first energy

---

[11] In the remaining part of this section, only the formation of Space and its properties will be discussed. The formation of time is analyzed in next sections. That's why the term "space-time" is not used here.

waves and particles etc. All these suggest big amounts of energy "entering" the universe. This is in line with the Double Existence Model. The first contact of CES and STK cause the first moment of energy projection from CES on STK and create ST.

Figure-12 shows a fictional example of complex energy wave function of an electron. In this example, it is assumed that two of the four wave components are projected on ST. Wave components 3 and 4 are not projected on ST and have no effect on the electron's mass but number 4 determines the spin. This is just an example to explain the logic of the model. The point is there are different wave types in CES which interact with ST in different ways depending on the complex wave function $f_{CE}(\vec{x})$ and how it is treated by $Q(\vec{a})$. So, the energy waves which form and span the space must be a specific type of complex energy waves with specific parameters. Instead of becoming matter in ST, they produce the building blocks of Space.

What are the main properties of those waves and what is their relation to ST? Figure-14 shows the interaction of a 1D energy wave interacting with a 2D space along x-axis, as an example. The energy wave applies a kind of cosmic pressure and creates "space". As shown symbolically, every period of the sine wave creates 1 unit length of space. When more periods of wave function apply -which means more energy is poured into ST- more space is created e.g., the universe expands. In other words, each package of specific amount of energy wave creates a specific amount of space.

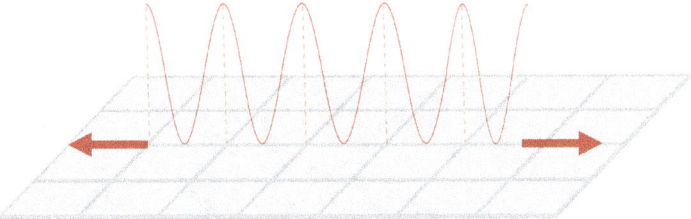

Figure-14. Creation of space or expansion of the universe by a specific type of Complex Energy projected on ST. Each period of a sine function creates a unit length of space along X-axis. The maximums of the sine function build "lines/borders" of the space unit (or the coordinate system) symbolically.

Figure-15 shows the same logic for a 2D sine wave, creating a 2D space on XY-Plane. 2D sine wave spans the space on XY-plane and each wave package creates a specific amount of "unit area" in 2D space. Let's use the popular term "pixel" to describe it. This analogy will help to simplify to explain the examples. Each pixel of the sample 2D space is created by a specific amount of Complex Energy. If we extend this 2D Space model to a 3D Space model like ours, it means voxels[12] are created.

---

[12] The Word pixel is derived from "Picture element" to describe the two-dimensional digital components of digital photos or videos. The Word "voxel" is the 3D version of it, derived from "volume element".

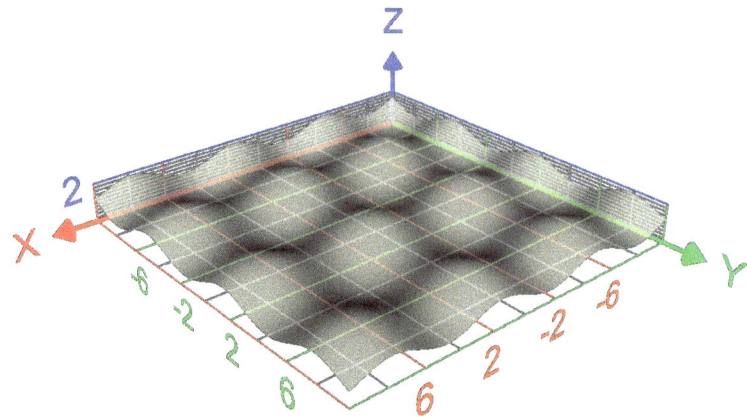

Figure-15. 2D sine wave example, creating a 2D Space along XY-Plane. The Complex Energy waves build/expand the 2D Space. The number of 2D sine packages define the amount of area i.e., the number of pixels created. If the periods of the sine functions are smaller, then the size of the unit area (one pixel) gets smaller. This means that the frequency of the Complex Energy determines the "resolution" of the universe (and minimum possible length and area.)

There are important results based on this analysis:

- The fabric of the space is built by energy packages of a certain type of Complex Energy waves. They build and expand the size of the space (or universe) by entering into the universe. Using the analogy with the communications systems, these Complex Energy waves are called the "**Carrier Waves**"[13] in the model.

---

[13] These waves build and span the space. Anything in the universe is kind of standing "on top of them" like the information signals carried by carrier waves in radio communications using a modulation technique.

- Their wave functions determine the fundamental geometrical structure of the space. In an empty space, the Carrier Waves build an orthogonal straight 3D[14] space (coordinate system) which can also be described as a set of cubic voxels.

- The size of these voxels is determined by the Carrier Wave function's period i.e., the size of the related energy package. In other words, the frequency of the Carrier Wave functions determines the resolution of the universe.

- This means; according to DEM, there must be a minimum size of distance, area, and volume possible in the universe which leads us to Planck Length.

- The Complex Energy Waves which are projected in the form of matter/energy will be referred as "**Material Waves**" in the model to distinguish from Carrier Waves.

- Carrier Waves must be very strong or dense signals, especially compared to the Material Waves, because their effect on ST is the dominant one. A big amount of mass (material waves) must be accumulated at one point of space to cancel out their effect and create singularity (a black hole).

- The frequency of the Carrier Waves must be very high compared to the Material Waves because the resolution they create in the universe must be higher than any object in the universe so that it can exist and

---

[14] Assuming the Space Time is 4D and space has only three spatial dimensions, by neglecting theories about higher dimensional universe models, at this point.

be observable. For example, the minimum possible length in the universe must be smaller than the size of an electron. Otherwise, it cannot contain the electron as it is. Thus, the electron field's frequency must be smaller than Carrier Wave's.

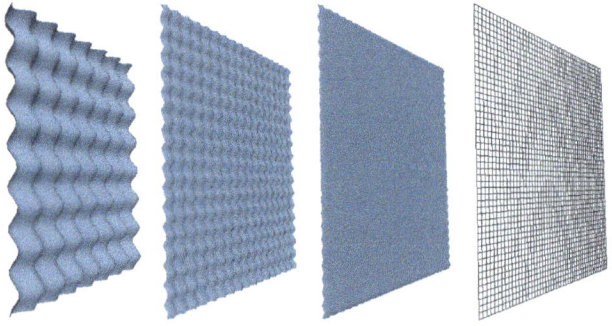

Figure-16. 2D Space example. On the left most, Strong Carrier waves constitute a 2D space. From left to right, the frequency of the Carrier Wave gets higher. As a result, on the right most one, a flat space is observed and in wireframe mode, it's like an orthogonal coordinate system with square pixels.

Figure-16 shows the summary of the creation of the Space. Strong Carrier Waves build the Space. If their power and frequency is homogenous, they constitute a flat 2D space which has identical pixels with same minimum length and area all over the universe.

76

In the real world, Carrier Waves span the 3D Space. Assuming there is no other matter and energy in the universe which would affect this structure, the Space has a perfect and homogenous orthogonal geometrical structure. The Carrier Waves constitute the fabric of the universe, create the spatial geometry, build the 3D space which is the landing field, home, or the trap of Material Waves.

U^U

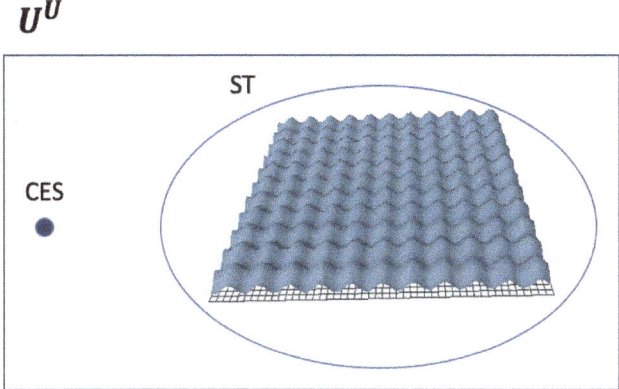

Figure-17. This new version of U^U presentation shows the interaction of Carrier Waves and ST. The 2D Carrier Waves are everywhere and they create the 2D Space. Their origin is CES. They are not projected in the form of energy or matter on ST, rather they built and span the ST at first place. Their complex functions differentiate from the Complex Energy Waves' functions.

By integrating this process to the model, we obtain an enhanced version of U^U presentation which includes more details about ST. This is shown in Figure-17. Carrier Signals are placed within the Venn diagram of ST. The whole space (or the universe) must be filled with Carrier Signals as shown in the figure so that the voxels are created, and the space exists. Any zone without Carrier Waves would cause deformation of the space like singularities.

There are two main scenarios about Carrier Waves' presence in the universe:

1. The amount of total Carrier Waves is constant. All of these were projected during the Big Bang or in a short period of time after the Big Bang. It still causes expansion of the space (universe) therefore, its energy density is getting lower in time, meaning, the expansion can get slower in time. In this case, the expansion of the universe is probably caused by the voxels getting bigger.

2. There is continuous flow of Carrier Signals into ST. In this case, the expansion of the universe is caused by this energy flow and the energy density of Carrier Signals is constant. So, the speed of the expansion is depending on the energy flow. In this scenario, the size of the voxels is constant and the expansion is caused by newly created voxels.

Some hybrid scenarios of 1 and 2 are also possible. For example, the energy flow of additional Carrier Waves may continue but voxels may be getting bigger anyway. Many similar combinations can be produced. Only some possible scenarios are listed here to present the logic of the mechanics.

Carrier Waves build the fabric of the Space and expand it. Unlike the Material Waves, they don't cause gravity. They are just the opposite in many ways. The way both wave types interact with ST are different because of the different natures of their Complex Wave functions, $f_{CE}(\vec{x})$. The way they are defined and described in the Dual Existence Model (DEM); they refer to Dark Energy as we call it officially.

Dark Energy is the projection of Carriers Waves on ST.	(H8)

The effect of Carrier Waves on space causes the space to exist and expand, based on their power distribution (and competition with Material Waves) in time. They interacted with STK in a singularity and created the Space.

There is an alternative scenario where if,
- There was nothing else which interacted with Carrier Waves at Big Bang, meaning STK was not any kind of object/energy, just an event which triggered all of this,
- There is no cosmic medium which interacts with Carrier Waves and forms the Space now.

In that case, it is also possible that the Space is just an energy field of Carrier Waves (projected in the forms of Dark Energy in observable universe) which hosts the Material Waves.

Figure-18 shows all the information and results of this section and presents the structure of U^U (in a 2D space format) as far as described.

In a summary,

- Big Bang is the first contact of CES and STK which created ST and initiated the energy projection on it.

- A specific type of complex energy waves in CES (Carrier Waves) created the Space from singularity and started to expand it.

- Carrier waves create a straight orthogonal 3D space and build cubic voxels which also means that the space is quantized.

- Based on the frequency (or unknown similarly affecting parameters in CES) of Carrier Waves, the geometry of the Space is determined and there must a limit for the minimum possible distance in Space (Planck Length). This distance directly/indirectly determines the size of the space-voxels.

- Expansion of the universe is caused by (most likely) new Carrier Waves being projected on ST.

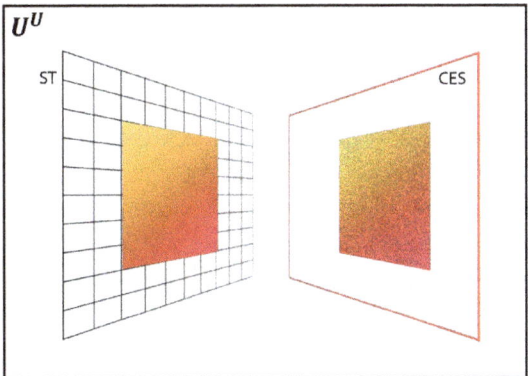

Figure-18. The final look of U^U containing the results of this section. Carrier Waves are projected on ST which constitute a uniform space shown in wireframe format. Some Material Waves are also projected which are shown in orange-to-red colored squares on both sides. These waves appear in matter or energy format when projected on ST.

4.2- What is Time?

Time can only be experienced if there is change. If everything was completely static with no change, we couldn't realize the existence of time (and we wouldn't exist at all). Figure-18 represents such a frozen moment. It's just an instance of time, one single moment where the Carrier Waves constitute the Space, and the Material Waves are projected on it. There is no time in the figure as it is.

There are two components if time and its effect on change must be analyzed according to the Double Existence Model:

1. The source of change, CES and the Complex Energy in it. As explained in previous sections, the Complex Waves are projected on ST and whenever they do so, they cause presence of matter/energy in compliance with their complex wave functions. Each time they repeat the process, they create a new state (another solution of their complex wave function $f_{CE}(\vec{x})$), a new reality in micro-level and create the change.

2. The Space hosts the complex energy waves in the form of matter/energy again and again.

To understand the mechanics of time, the relation of these two steps must be analyzed. Figure-19 shows two instances of time and how the change is represented along the interaction of CES and ST.

Complex waves in CES environment may cause change in ST environment because of two different reasons:

1. There is a change in Complex Energy Wave in CES environment which results a change in ST. This is shown in Figure-19 at the bottom right corner part of the square which becomes yellow(light) at t_1 in both ST and CES.

2. There is no change in CES environment but the re-projection of the Complex Energy Wave on ST has a different result in ST environment in compliance with its wave function $f_{CE}(\vec{x})$. This is shown in Figure-19 at the top left corner part of the square which becomes yellow(light) at t_1 in ST only.

This process can be considered as a state-machine. At each moment, some complex energy waves are projected on ST and one probable solution of their wave functions $f_{CE}(\vec{x})$ is realized. So, every particle or energy wave in ST jumps to their next state.

A useful analogy is the system clocks in digital circuits. Each time the system clock generates the trigger signal, each component in the circuit implements one task like a step of a mathematical operation, writing one bit to memory or reading one bit from a disc etc. All components accomplish their job simultaneously and wait for the next trigger signal to implement their next task. At the end of each cycle, the states in the system are changed (data in the memory is updated, some new content is displayed, some new commands get ready for action etc.).

$$t_0$$

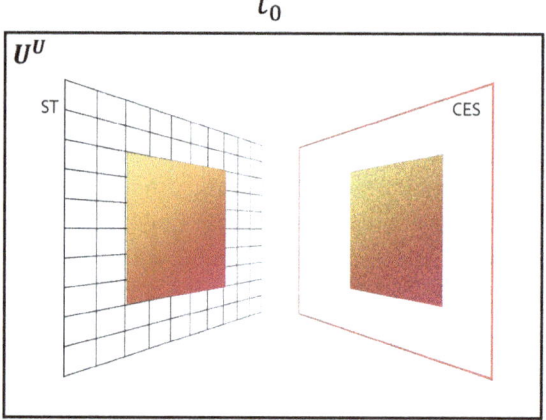

$$t_1 = t_0 + dt$$

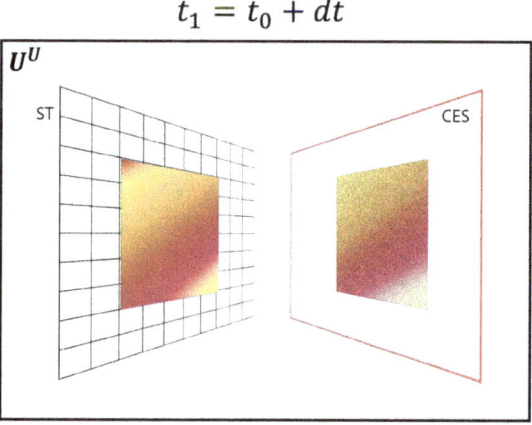

Figure-14. CES and ST in U^U environment at two different points of time t_0 and t_1. At time point t_1, there are two changes in ST environment. The upper left corner in red(dark) is the new state/solution of the complex energy wave function in CES. The bottom right corner in yellow (light) which is caused by a change in CES domain.

84

There must be such a mechanism in ST-CES relationship which triggers the re-projection of Complex Energy on ST, create change in the universe. The universe jumps from one state to the next one, by realization of all $f_{CE}(\vec{x})$ functions of projected Material Waves for that instance of time. Defining time is about to describe this trigger of universe's state-machine.

The conventional perception of space and time relation is not like a state machine or digital circuits. It is a more "analog" one than the described one above. Based on the assumption that all the energy and matter is present in ST, their change is seen like an analog process continuing in time. The Space component of ST is one and single container of matter & energy, and the time component is responsible for the change in it.

It is not possible to define the real relation of change and time based on this perspective because they are described just as naturally related processes. For example, when a rocket is launched, time flows along with it by itself during the flight. There is no connection which makes time to flow and the rocket fly. We assume time is always there for us, flowing in parallel to everything happening.

Of course, time continues to flow by itself, it's not wrong but the mechanics is different. DEM suggests that Complex Energy Waves are projected on ST consecutively which results the change. This takes us to a different perspective to define the origin and mechanics of time. Figure-10 shows the change of an electron's position at two different time points. In the figure, the electron wave is symbolically assumed to hover in CES and move towards two time points and contact ST twice. Combining this logic with examples in Figure-14, an updated version of U^U is obtained which includes the scope of time. This version of U^U presentation is shown in Figure-15.

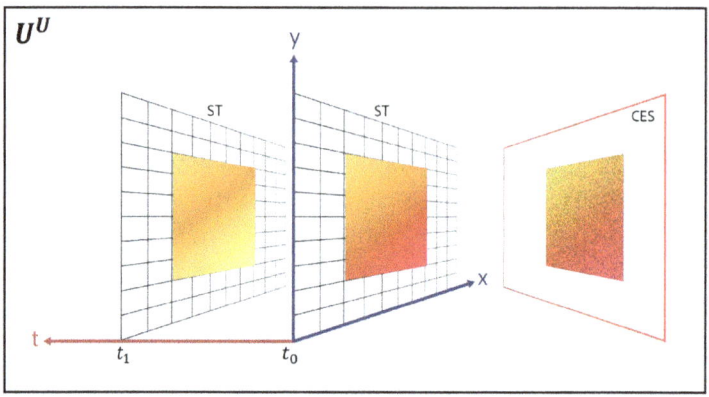

Figure-15. U^U representation including time. The waves in CES are projected twice and resulted two different states in ST. The waves must "move" from t_0 to t_1 to be projected in that moment and reappear in ST. A new moment is created with new states of matter and energy in the universe.

Time is a dimension according to our mathematical modelling. It's created with Bing Bang like the Space was. As explained in the previous section, Space is created by so called, Carrier Waves. Considering the fabric of space time is whole, Carrier Waves must also be the source of the time dimension but as time is not a spatial dimension, its existence must depend on a different parameter/property of Carrier Waves. A different kind of interaction must have caused the time dimension which is nothing like spatial dimensions that you can choose a direction and move. It has no directions or geometry but it triggers and causes the change in Space.

86

In Figure-15, two time points are shown. The Complex Waves in CES are projected two times on ST resulting two different states. The second one (t_1) represents the next moment of the first one (t_0). Avoiding any geometrical or spatial meaning, the waves in CES must "go" from the first moment to the second one in time dimension to reappear at t_1. This will trigger all the matter & energy in ST to go to the state at the next moment, like the memory and processors in digital circuits.

Let's go back to the "movie universe" analogy. Consider a movie and its characters represent a universe, where all the events happen in the movie. This universe is displayed on the white screen in a movie theater. A conventional projector with a roll of chemical film is used to display the movie.

In this case, every moment of the movie universe, i.e., every frame of the movie is already there. It's printed on the chemical film. If anyone would unroll the film and look at the frames, he could access any frame — any time point in the Movie universe- arbitrarily. But the characters in the movie would live the events in a sequence only when they are projected on the white screen. Each moment would be created with the projection of that moment's signals (information about the state of that moment) on the white screen. This would happen when the related frame is placed in the optical system of the camera in front of the bulb and projected.

In a kind of universe, it is possible to stop, rewind or play forward the time. All the historical information is recorded and accessible by an external observer. The people living in that universe and the characters of the movie will experience the same events without even noticing if the observer rewinds the movie. They can't change anything because everything is recorded. No individual time travel is possible which would mean a person frees himself from the flow of events and be somewhere different other than he is supposed to be.

The main characteristics of the "movie universe" are helpful to understand and play with the ideas like time travel or free will. It also presents some references for the Double Existence Model. According to DEM, there are some similarities with the movie universe, but some flows are just the opposite. Everything about the movie universe is related to display a recorded movie and the mechanics are based on this.

According to DEM, **Space Time is like a recording camera** and its chemical film. **CES is the source to be recorded**. Some of the complex energy waves (like the photons arriving to the film in the camera) in CES contact the Space Time and are projected on it. This means that the real events happen in CES, the origin of matter and energy (the cast of the movie) are in CES. Space Time is the recording and simultaneously displaying medium of all this.

ST holds/hosts/traps the energy waves and they appear in the form of energy or matter in our universe. This is the creation of one moment in space-time. To create/record the next moment, a new empty frame of film is needed in cameras. This is also what happens during the interaction of ST and CES. As shown in Figure-15, at a new instance of time t_1, there is an instance of empty space where the complex waves can be projected/recorded again.

Figure-16 shows the generalized version the of space-time in Figure-15 where multiple instances of time points and the related empty spaces are presented. **For each moment in the universe, there is such an empty 3D Space** ready to host and trap the Complex Energy Waves from CES. When the Material Waves arrive to a new instance of space and they are projected, then next states of particles & energy waves in the Space-time are created which is the next moment in the universe.

To do that, the Material Waves "move" from one instance of space to the next which means going from one moment to the next. Despite the word "move" is used, this is not a spatial or geometrical behavior. This is a movement in reference to space-time beyond our geometrical perception.

Carrier Waves create the 2D spaces in Figure-16 (3D Space in our real world) as explained previously and additionally, they create copies of these 2D spaces along the time dimension. The movement of Material Waves is along this dimension t, which is shown in Figure-16 as a horizontal coordinate axis (from right to left) for presentational purposes. In fact, it is not such a geometric attribute with no spatial direction.

In summary, the Carrier Waves constitute the Space Time where they span the 3D Space as we know and they create multiple instances of it where the Material Waves can be projected again. So, new states of matter & energy can be produced in new empty universes at new moments. Parallel to this, all the complex energy waves, the source, continue to exist in CES and their Complex Energy Functions are valid. The projection is one random result of their functions with the boundary conditions set by ST as feedback.

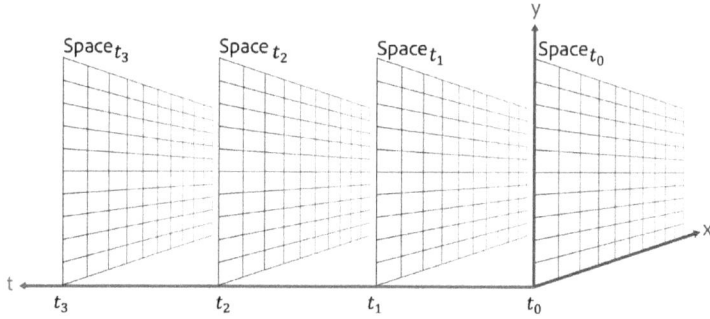

Figure-16. Presentation of Space-time with multiple instances of time and space. Each moment of time in the universe happens by the projection of Complex Energy on the next instance of space which has the new set of states, the reality of next moment.

Combining all the information and logic presented in this section, we can define time as:

> There are empty Spaces for every moment in the universe. Time is the creation of next reality state by Material Waves being projected on those empty spaces, sequentially.

(H9)

Figure-17 shows the full version of U^U according to DEM, including all the components CES, space-time and different moments/states of projected matter & energy. Each moment, the Complex Wave Functions of projected Material Waves, create results in ST domain which is shown with colored

90

rectangles. The results of Wave Functions $f_{CE}(\vec{x})$ at that moment which are also based on the initial (or boundary) conditions and feedback from ST of the previous moment, determine the states in ST. The link of initial/boundary conditions between two states or moments enables the causality in macro level where $f_{CE}(\vec{x})$ has probabilistic projections/results in micro level.

This mechanism makes the time flow into the future, moment by moment, which also means that the universe is a discrete time system with a very high clock frequency.

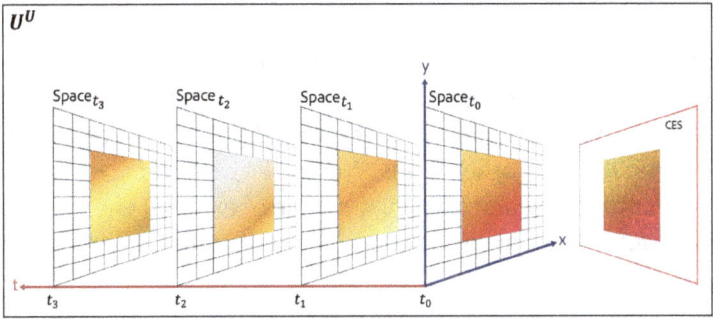

Figure-17. The most generalized structure of U^U according to DEM. There is one source of complex energy, CES. Carrier Waves from CES create and expand the Space in 4D Spacetime with multiple instances (wireframe rectangles). Some Material Waves arrive at ST and are projected on it. They create matter & energy in ST (Colored rectangles). Their projection may have the same or different solution/projection each moment. There is also a causal connection of their solutions/projections because of initial conditions created by ST and fed back to CES.

4.3- How much future do we have?

Figure-17 shows how the universe is structured in U^U according to DEM. This is a complete picture, modelling existence of matter, energy, and time in the universe. The next important problem to analyze is about the amounts or sizes of components in Figure-17. This is essential to understand the reality of our existence.

The size of CES is irrelevant because DEM is defining CES only as the external source of matter and energy in Space-time. Its size, geometry and number of its dimensions (if there is any) are unknown and irrelevant.

The number of CES must be one. If there were more than one energy source feeding ST, then there would be a chaos because of superimposed waves from different sources which interfered each other. Thus, the order and stable causality lead us to the idea of one single source of external energy. This is important because it has an important effect on what our reality is.

More interesting discussions can be made about the size and number of empty spaces in space-time. It refers to the size and number of wireframed rectangles in Figure-17. As we know, the universe was a singularity when Big Bang occurred and it started to expand. This is the expansion of 4D space-time and it should mean that more and more "space and time" is produced along with it. It is possible to physically observe and see the size in spatial dimensions, the 3D Space. According to the observations and calculations, if we accept the universe is roughly sphere shaped, it seems to have a radius of 46 billion light years. That means we have a pretty good idea about the size of the Space, the wireframed rectangles in Figure-17.

What about the number of them? The expansion of space-time from a singularity seems promising about creating a lot of time while creating 92 billion light years big Space. We already are aware of 15 billion years history of the universe which is a long-time frame. Still, there is an important question to ask:

"How much future do we have?" or "How far[15] time do we have ahead?".

We can look at the sky and see the stars which are millions of lightyears away. We know, we can go to any of them if we have the right spaceship, because the space between us and that star literally exists. That star is standing there at that distance. It is possible to observe that distance in spatial dimensions. Can we do the same thing, observing time?

To understand and answer these questions, the scenarios about the existence of empty spaces of moments must be analyzed. Figure-18 shows the components of all possible scenarios.

---

[15] The correct English phrase is, of course, "How much time…". I used the Word "far" intentionally to create the perception of time as a dimension which is correlated with a kind of "distance" on its own way. We are moving towards the future which "stays" ahead of us.

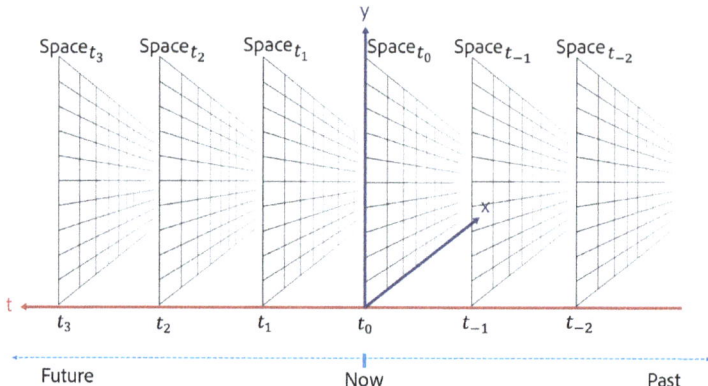

Figure-18. The set of past, present and future empty spaces. The existence of $Space_{t_0}$ is the only certain scenario. Past and future spaces may exist as they are shown in the figure but it is equally possible, some or all of them are not there "all the time".

Theoretically, the combination of the following three states is possible:

1. The present time, $Space_{t_0}$, exists for sure and there is only one piece of it. Otherwise, the universe wouldn't exist now.

2. The spaces of past ($Space_{t_{-1}}$, $Space_{t_{-2}}$,...) existed for sure when they were the host of the past but there are two kinds of uncertainties about them.
   a. Do they still exist?
   b. If yes, how many of them still exist?

94

3. Do the spaces of future ($Space_{t_1}$, $Space_{t_2}$,...) exist "now" before it's their turn or are they created by the Carrier Waves "just in time"? If they are already there and waiting for their turn, then how many of them are there waiting for us? Does the production continue?

Some combinations of 1,2 and 3 create interesting scenarios about our existence and its reality.

1. 4D expansion of ST is valid both for space and time and there is a room for billions of years of time created (like the space, 92 billion lightyears wide). This seems to be the most probable scenario because the assumption is based on the analogy to spatial expansion, we already observe. This scenario is just an extension of the spatial expansion in time dimension. In this case, **all the Space instances shown in Figure-18 exist**. Resulting:

   a. There is a first Space in the past which is created at the Big Bang, so the past has a finite number of Spaces (which contain the 15 billion years of history of the universe)

   b. There is no information about the number of future Spaces. The expansion of the 4D ST may have created billions of years but it is not possible to determine how much time is left. We can't know how far the Carrier Waves have gone in time dimension and created us room to go further into the future.

   c. Another question would be, "If the expansion in time continues (3D Space continues to

95

expand as we know), "is the production faster than consumption?" We may consume time faster than the Carrier Waves produce it. Maybe, the production was faster at the beginning and we didn't hit the wall but now, we may be approaching to it.

d. This scenario suggests that we have a finite amount of time ahead which is already created. However, we don't know how much.

e. The historical information is preserved in "past Spaces". It is theoretically possible to access that information.

2. Another interesting scenario would be the case, where **no past and future Spaces exist**. In this case, Carrier Waves only span the 3D Space but don't add up new amounts of time in time dimension. They only create the empty Space of current moment after they use the previous one. This would be a kind of flashing universe. The new space is created by Carrier Waves and right after that (or simultaneously) Material Waves are projected on it and the present moment is built (and disappears afterwards). Remarks:

a. The only Space which really exists in Figure-18 is $Space_{t_0}$.

b. The next moment is produced "just in time" by the Carrier Waves. This means, we don't have certain amount of future ahead. Every moment is created one by one without any "spare future".

c. There is no single free space ready for the next moment. Thus, if somehow the process breaks, our existence will be terminated instantaneously.

d. As there are no past spaces, all the historic data is lost. Thus, time travel to the past is totally impossible. There is no accessible past at all.

3. There are no future spaces but all past Spaces exist. In this case, the historic data is preserved and accessible. There is no guaranteed amount of future ahead. The Carrier Waves instantaneously create the future adding new moments (spaces) to the universal history.

4. There are no past spaces but future Spaces exist. Certain amount of future is guaranteed but the historical data is lost.

5. There are finite numbers of short-term future and past Spaces, which would mean some amount of future is guaranteed and only some historical data is preserved. As an example, assume that 5 days of spaces do exist in both directions. So, the universe exists as a 10-day long story. This kind of universe would be like a signal fading in into the future and fading out from the past, continuously.

In summary, despite our perception about time being an eternal source or infinite long dimension in ST, we don't know how much future we do have ahead. Like the universe has a finite radius which expands time dimension may have a limit, too.

4.4- Special Relativity and the Uniqueness of Reality

Having defined what time is and how it works, we are ready to discuss the speed of it. Special Relativity suggested so many new and interesting rules about how photons and matter move in space and how the flow of time is affected by this.

Photons always move at the speed of light, c, and they don't experience time. Objects with mass can never go up to that speed and as much as they get faster, time gets slower for them. It is not easy to understand what this means without dealing with mathematics and see how these parameters change. Even then, it may be difficult to imagine in practice, how time can flow differently at different speeds.

Another strange fact to practically comprehend is that every object moves at the speed of light in space-time. This speed is not defined in space, it is defined in space-time. So, it is about a movement in a kind of hybrid coordinate system with spatial and time-based axes. The first question to ask would be: "Why is it equal to c?", but the more important question is: "What is the source of this speed?". The link between an object's spatial speed and the temporal speed makes these questions even more complicated. The total speed at "c" in space-time is a kind of resultant vector of spatial and time-based speeds. Mathematically, this can be shown easily by using Special Relativity equations, but what kind of universe does that? What mechanics causes the speed in space-time? How is it divided into two components?

This section focuses to analyze and explain all the phenomena about the direction and speed of time, time dilation, why every object moves at "c" in ST. This will enable the necessary set of information and deductions to describe the nature of our reality and the uniqueness of it.

Let's start by looking at the ST coordinate system in Figure-18 using another perspective, from top, where we only see a 2D projection of it with x and t axes (Figure-19). This 2D version can represent the relation between time, space, moving particles/waves and complex waves which create them. The empty space planes $Space_{t_0}$, $Space_{t_1}$,... in Figure-18 are horizontal lines, parallel to x-axis in the graph.

In Figure-19, two types of Material Waves' projections are shown. The first one is a Photon Field which is shown as a vertical (red) wave, shown parallel to time-axis on the left. Photons don't experience time and move at the speed of light. This means that they don't move in time dimension. Thus, they must be projected on every moment at once. Otherwise, they would stick just in a moment of time, and we wouldn't observe any light beam coming from any source.

Thus, the Material Waves in CES which create the Photon Fields (MW_p) must move in a parallel direction to the x-axis[16] and vertical to t-axis. This means that the photons don't move in time and their full "kinetic energy" is used in 3D space, resulting their speed, "c". In other words, their movement in 4D ST at the speed of light is fully projected to 3D space at the same speed without any time component. That's why photons cannot stop. As long as they exist (and not absorbed by other particles), they move at the speed of light.

---

[16] This is valid for the sample 2D Space Time in Figure-19. In real world, it must be projected in a parallel direction to our 3D Space.

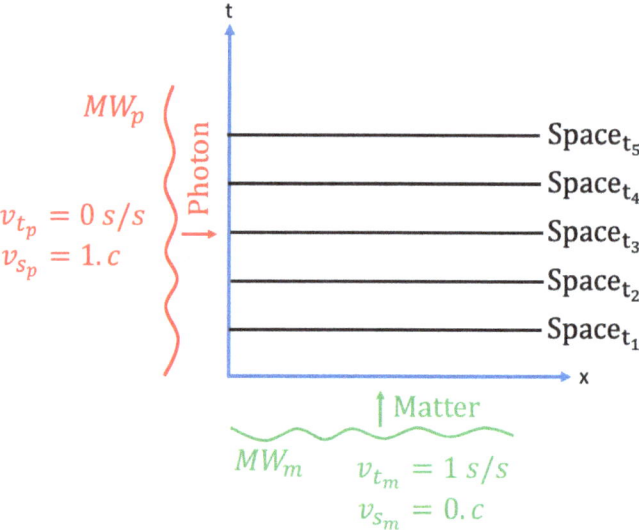

Figure-19. The 2D representation of space-time and the vectors of complex energy waves creating Photons and Matter in ST. These two wave types are orthogonal to each other which differentiates their behavior in ST. MW_m waves move and are projected on ST in a parallel direction to time which causes objects to experience time at the speed of 1 second/second if they are at rest. Likewise, MW_p waves are parallel to 3D Space, and they move at the speed of light, c.

The second Material Wave (the horizontal wave, parallel to x-axis, in green color) in the figure is projected not only as an object, but also as a matter[17]. As an example, it can be a pencil. A pencil is an object with mass, it's a real matter. That means time flows for it as it stands still. Isaac Asimov answered the

---

[17] This wave is assumed to project a random object on ST rather than a specific particle like an electron, to have a generic example.

question "Is time travel possible?" in a very basic and elegant way in one of his books. He wrote: "We do travel in time, at the speed of one second/second, forward, into the future.". That's the case for the pencil. It travels in time at one second/second. As the photons can't stop in Space, the pencil can't stop in time and moves continuously into the future. On the other hand, it can be at any random place in Space, and can move to anywhere. That's why, the Material Waves which create matter on ST (MW_m) must move orthogonal[18] to the x-axis (Space) and parallel to t-axis.

This is what differentiates photons and objects (particles with mass). Photons and objects[19] are originated/projected by two orthogonal[20] complex energy wave types in CES. Even if the object is in motion in 3D Space, its waves, (MW_m), always have a component parallel to t-axis which is orthogonal to MW_p. That's why objects (matter) cannot go up to the speed of light. The orthogonality of these two waves (MW_p and MW_m) also determines their attributes of having mass.

In summary, the directions of projections of two Complex Energy Wave types on space-time determine how they appear and move in it. The mass, spatial velocity and speed in time of their projections are different and not convertible to each other by applying mechanical effects within ST. They have two types of speed (movement), as shown in Figure-19, v_s, the speed in the space and v_t, the speed of time. v_s and v_t of Photons and Matter are shown in the figure separately where v_s is given in the form of a multiple of "c" to show the

---

[18] if the object is in rest in 3D Space
[19] This is just a generalized expression to symbolize matter. In fact, this is true for any particle with mass and its energy field.
[20] if the object is in rest in 3D Space

orthogonality between two speed types. We can show this by presenting them in vector format, as in (9).

$$\vec{v_p} = \begin{bmatrix} v_{t_p} \\ v_{s_p} \end{bmatrix} = \begin{bmatrix} 0 \\ 1.c \end{bmatrix} \qquad \vec{v_m} = \begin{bmatrix} v_{t_m} \\ v_{s_m} \end{bmatrix} = \begin{bmatrix} 1 \\ 0 \end{bmatrix} \tag{9}$$

The relativistic equation for time dilation is shown in (10), where v_s is the spatial speed of an object, t is the amount of time while object is still and t' is the amount of time at the speed of v_s.

$$t' = t. \sqrt{1 - \frac{v_s^2}{c^2}} \tag{10}$$

Because the amount of time passing in (10) and the speed of time at the same conditions must have a linear relation, we can replace the variables about time with variables about speed of time. This leads us to (11) using the notation in Figure-19.

$$v_{t_m} = v_{t_{m0}}. \sqrt{1 - \frac{v_{s_m}^2}{c^2}} \tag{11}$$

102

As shown in Figure-19, $v_{t_{mo}}$, the speed of time for a still object is one. Putting this value in (11) and reformatting it, results (12) which is a known formulation.

$$c^2 = v_{t_m}^2 c^2 + v_{s_m}^2 \qquad (12)$$

(12) is one of the mathematical expressions, stating that, all the objects move at the speed of c in Space Time. It shows the relation of spatial and time-based speeds and how they relate to the speed of light.

Simplifying (12) by c^2 and using the format for v_{s_m} in Figure-19 as a coefficient of c, $v_{s_m} = a.c$,

$$1 = v_{t_m}^2 + a^2 \qquad (13)$$

"a" is defined as the ratio of the object's speed to the speed of light. Its value can be between zero and one. On the other hand, v_{t_m} is the speed of time, and its value range is the same as "a". It is the speed of time for the object but because its full value is "1", it can also be interpreted as a ratio to the full speed of time. So, both variables can be considered as ratios or percentages of maximum possible values.

Both (12) and (13) represent circular functions. Figure-20 shows the graphics of (13). This is a known representation of time dilation according to the Special Relativity. It shows the temporal speed's relation with spatial speed and how time goes slower if an object gets faster in space.

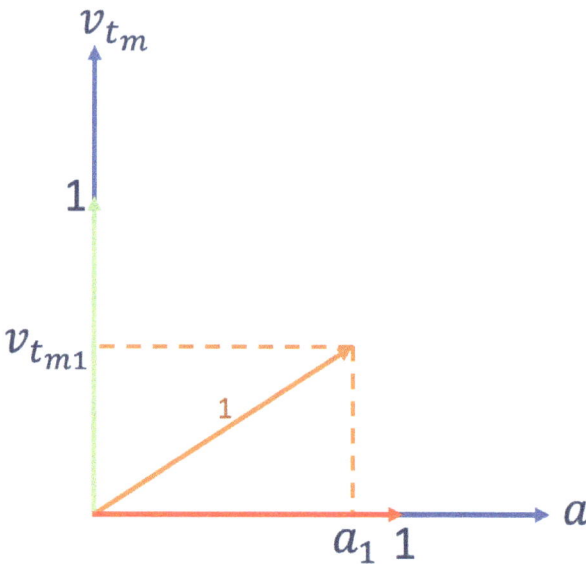

Figure-20. The relation of spatial speed and temporal speed. The vertical/green arrow belongs to the green MW_m wave shown in Figure-19. It's a still standing object which experiences time in full speed. The horizontal/red arrow belongs to the red MW_p wave in Figure-19. Its projection on ST is a photon which doesn't experience time but can travel in space at "c". The orange arrow belongs to an object at a specific speed of $v_{sm1} = a_1 . c$. It experiences time dilation. The speed of time for that object is v_{tm1}. The radius of the circle is one which represents the speed of light, "c". The figure shows that every type of object/wave has the speed "c" in Space Time which has two components that we can directly observe.

The Double Existing Model suggests that the waves which create matter and energy in the universe are projections from CES. So, the red and green waves in Figure-19 are Complex Energy Waves from CES being projected on ST. In other words, the red and green arrows in Figure-20 are vectors which represent the direction of those waves' projections. Obviously, the sizes of the vectors are correlated with the speed of the waves and of their projection relative to the space-time.

Combining all this info and adding the fact that the resultant vector always has the size of "1" which represents "c", speed of the light, the following hypotheses can be defined.

The constant speed (which is "c") of every object in space-time is caused by the relative movement of CES with respect to the Space Time.   (H10)

This movement causes the projections of complex waves to move at any direction at the speed $v_{CEW} = c$ in space-time.

> The existence and values of speed of the light "c" and the speed of time $v_{t_m} = 1$ second/second are caused and determined by the movement of complex waves at the speed of $v_{CEW} = c$ with respect to Space Time.
>
> Thus, speed of light and speed of time are two orthogonal projections of the same phenomena, the movement of the Complex Energy Waves at $v_{CEW} = c$, which links them and causes Time Dilation.

(H11)

(H10) and (H11) define the proposals of DEM which explain the fundamentals of the relation of movements in space and time. The main root cause of change in the universe, creation of time and the value of "c" is the relative movement of CES (thus, Complex Waves and their projections on ST) with respect to the space-time at a certain speed[21].

We can describe this by using a simple model. Let's assume that CES is a moving box outside the Space Time, at the speed of $v_{CEW} = c$. There are many types of energy waves in CES-Box and they move into different directions. Thus, the direction of their projection on ST is a resultant vector of their movement in CES and the velocity vector of CES. This causes

---

[21] Because no geometry or medium is defined in U^U environment, this is not a spatial speed as we understand. This could be rather a difference of parameters of Material and Carrier Waves related to propagation characteristics.

some of them being projected parallel to the temporal dimension (green waves and arrow in Figure-19 and Figure-20). Some of them are vertical to the temporal dimension (red waves and arrow in Figure-19 and Figure-20). Some of them belong to the moving objects in Space Time and have components both in temporal and spatial dimensions (orange arrow in Figure-20). Their projections may have different directions but the speed of the projections is always $v_{CEW} = c$. Such an example is a simple model, explaining (H10).

(H11) suggests that the speed of light and speed of time are created by the same root cause. In a way, they are the same thing in the origin. The only thing which differentiates them is the direction of the projections between CES and ST. The angle of the direction determines the size of the components of $v_{CEW} = c$ in Space Time which sets the speed of time vs. spatial speed for that object. There are three scenarios as shown in Figure-20:

1. For a material wave, projected vertical to 3D space that creates a still object/matter in Space Time, $v_{CEW} = c$ projects only a temporal component at the speed $v_{t_m} = 1$ second/second.

2. For a material wave, projected parallel to 3D space that creates photons in Space Time, $v_{CEW}=c$ projects only a spatial component at the speed $v_{s_p} = c$.

3. For a material wave of a moving object in the universe, $v_{CEW} = c$ is not parallel to temporal or spatial axes and projects both v_{s_m} and v_{t_m} components on Space Time which causes $v_{t_m} < 1$, Time Dilation.

In other words, the most important constant in the universe, the speed of light, is the projection of the speed of CES/Complex Energy Waves with respect to space-time. If that speed of the projection is fully applied into the direction of time, it creates the temporal speed of 1 second/second. To analyze this further and see how time dilation occurs according to DEM, we can superimpose Figure-19 and Figure-20 and put two objects to compare their mechanics.

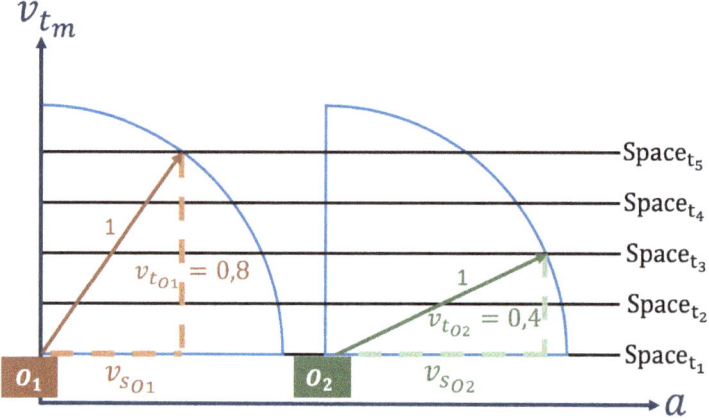

Figure-21. Space instances, which constitute the Space Time are superimposed with the ST coordinate system. The Objects O_1 and O_2 move at high speeds in space and experience time dilation at different levels. O_1 has the double temporal speed of O_2 which results it encounters with a higher number of empty space instances of new moments.

108

This is shown in Figure-21. The space instances $Space_{t_1}$, $Space_{t_2}$,..., $Space_{t_5}$ which constitute the Space Time in Figure-19, are put into the Space Time coordinates as in Figure-20. Additionally, two objects, O_1 and O_2, moving at different spatial speeds v_{so_1} and v_{so_2} respectively, are located. As they move at high speeds, comparable to c, they experience time dilation. As shown in Figure-21, the speed of time for O_1 is $v_{to_1} = 0,8$ where it's $v_{to_2} = 0,4$ for O_2. Both objects move at the speed of light, c, in Space Time which is shown with the normalized radial vector "1".

Figure-21 contains a very important perspective to understand the nature of time dilation and the reality of our existence. There are four discussion points about this figure which deliver the components of the hypotheses, proposed by DEM.

1. The temporal dimension is not a spatial type of dimension, it has no geometric properties. Figure-21 shows the speed of time as a vertical dimension, upwards and each new instance of space is lined up from bottom to the top in a sequence using a geometric order. This is only to show them on the paper because it is the only way to sketch anything about time where we can show some relation and make analogies. In fact, the temporal speed and the positioning of space instances have no geometric properties or relations as we understand, in the U^U environment. Although, a specific type of speed into the direction of time and a specific type of sequence of empty space instances exist between the Complex Wave Projections and the Space Time.

2. This requires a very important clarification about our understanding of time as a dimension and coordinate.

   a. If the objects O_1 and O_2 were two moving cars (starting at the same point and moving to the same direction) at the speeds of 100km/h and 150km/h respectively and we wanted to calculate their distance in an hour, we'd find it as 50km. This is obvious because O_2 is faster in space and can go further then O_1 can. If the movement is in spatial domain, geometric rules apply and the faster object ends up at a farther point than the slower object will do. This is valid for the objects O_1 and O_2 in Figure-21, too. As seen in the figure, O_2 is faster in spatial coordinates than O_1, $v_{s_{O2}} > v_{s_{O1}}$. If they start to move together side by side at the same point and into the same direction, they will never be next to each other again. They will be at different positions/coordinates in 3D space.

   b. This is not true for time. Let's take the twin astronaut example. A spaceship is launched in the year 2000 where the twins are 30 years old. John stays on Earth and his astronaut twin Tom leaves him behind. 50 years later, in 2050, John will be 80 years old. Tom will travel at a very high speed in the spaceship, experienced time dilation and only 10 years will pass according to him. So, when he returns, he is 40 years old. The twins travelled at different speeds in time along the process but, unlike the car example above, they will end up at the same year, same coordinate in time, in 2050. In other words, both start in year 2000, John travels 50 years into the future and Tom travelled 10 years into the future, but they are both in 2050. If John walks 50 meters and Tom walks 10 meters into the same direction in the backyard of their home, there would be 40 meters distance between them. That's not the case about time.

c. In Figure-21, the projection vector of O_1 has reached $Space_{t_5}$ where O_2 has reached $Space_{t_3}$. This may seem normal, because O_1 has the double temporal speed of O_2, it made four jumps (projections) through the space instances where O_2 has only two jumps. On the other hand, as explained in 2-b, they must be at the same moment, sharing the same present. Does it mean, they both must be at $Space_{t_5}$? The answer is clearly "yes". Otherwise, they are not at the same point of time. So, the "distance" they reached in the direction of time and the point of present they are staying at, must be independent of the number of jumps (projections) they make. In other words, the object's **temporal speed doesn't determine how far it goes in time, it determines how much it is affected by time**.

3. As explained in item 1, temporal dimension is not geometric dimension and space instances are not lined up in a row as shown in Figure-21. This is just a representation on the paper to show the logic. The notes in item (2) makes it clear though, the presentation in the figure gives us a hint about the

111

mechanism of time. The amount of the temporal speed v_{t_m}, which is the temporal component of an object's constant speed at "c" in Space Time, determines the number of space instances, the object's complex energy waves are projected on. As explained using (H9), time is the creation of next state of the universe, which a kind of state machine. So, the **"temporal speed of an object" is related to the number of state updates** of that object's complex waves (i.e., number of projections) on new space instances.

4. CES is the only and unique source of the Complex Energy Waves. As explained in item 3, its waves are projected on ST in different rates, creating different temporal speeds for those waves' projections in ST. The experience in Space Time is also unique as explained in 2b. On the other hand, the mechanics in Space Time effect the Complex Energy Waves in the moments they are projected on ST and create feedback to CES environment. As a result, the complex waves of CES have only one valid state for any given time point (presence) in ST which is uniquely related on the complex wave functions in CES. So, CES and its projection on Space Time define and constitute one single reality of presence.

These four discussion points lead us to an additional set of hypotheses which explain the reason of time dilation and the uniqueness of our reality and presence.

The temporal speed is related to the number of projections of complex energy waves on ST.

If an object moves in Space, its temporal speed v_{t_m} gets smaller than 1. This results the complex waves of the object have a smaller number of projections on new space instances. This decreases the number of state-updates and causes Time Dilation.

(H12)

The consistency of Complex Wave Functions in CES and their projection on Space Time define a unique reality. CES contains all the information about the unique reality of presence where Space Time is the medium which displays it.

Thus, the presence we experience, is the real presence and the only live moment in the universe.

(H13)

The uniqueness contained in CES and the reality of the presence in Space Time conclude that,

- The future does not exist in CES yet. No information is already created about the future, and nothing is accessible about it. Time travel into the future is impossible.

- The past has disappeared in CES, it cannot be experienced again or changed. It can only be observed only if the previous space instances still exist in U^U. Time travel into the past is impossible.

(H14)

4.5- Mass and the Space

So far, we have defined;

- the Carrier Waves, source of dark energy, which span and constitute the Space Time,
- the Material Waves, source of matter and energy,
- the projection of Material Waves on ST, which causes our existence in the universe,
- the source of time and speed of light: the relative speed between CES and ST,
- the definition and mechanics of time which works like a state machine and is affected by the angle between the Material Waves and ST.

All these define a set of rules about the interaction of space-time and the Complex Waves in CES. We could also use the term "Carrier Waves" instead of ST because, as explained in section 4.1 (and shown in figures 14, 15, 16 and 17), Carrier Waves constitute the "fabric" of space-time.

So, all the games between Material and Carrier waves, create the physical universe, the matter, the energy, time, time dilation and everything.

Let's move forward one step further and focus on the specific case, when the Material Waves are projected as matter with mass. What is the mechanics in that case, how are the Carrier Waves, i.e., the Space-time affected according to DEM so that:

- mass bends the Space-time.
- mass causes time dilation

As explained in 4.1, space is created by Carrier Waves which span it and form a straight geometry if the space is empty. This

114

shown in Figure-22, on left. The empty Space is straight and has the form of a classic orthogonal coordinate system. To make an analogy, the energy of Carrier Waves applies a kind of cosmic pressure, or they act like coil springs which generate the distance between the corners of the squares in the Figure-22. Because the figure is shown in 2D, this way they create a flat plane of squares. If the figure represents the micro world showing the minimum sized squares in the universe, these are the "pixels" of the universe, or in real life, in 3D Space, we can say they are the voxels of the universe. In this case, the sides of these squares (or cubes in 3D) are the size of Planck Length.

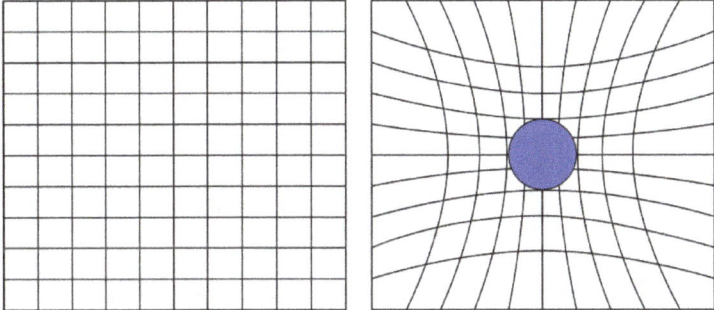

Figure-22. Empty straight Space (left) and Space bent by mass (right).

The real impact of mass on Space-time is shown in Figure-22, on the right. Let's assume, the blue circle represents the Earth. The bent lines around the circle show exactly how the mass of Earth affects the space around it. All the straight lines on the left image are reshaped. This was one of the most important outcomes of General Relativity. Einstein's theory suggests that

115

there is no gravitational force between objects with mass. Objects with mass bend the space around them and this creates the perception of the gravity.

Because the space itself is bent, it means the geometry of space is modified. Therefore, it is not necessary to have mass to be affected by gravity. This is shown in Figure-23. The red arrow represents a light beam moving from left to right. On the left, where the Space is empty, it follows a straight line. If the Earth comes into play, the light beam still follows the same "coordinate line" but as it's a curve now, the light beam's course is not straight anymore. That was one of the most beautiful propositions of Einstein. Photons have no mass but are affected by gravity. It was proven in 1919 by Sir Arthur Stanley Edington during a solar eclipse. That was one of the most solid proofs of General Relativity.

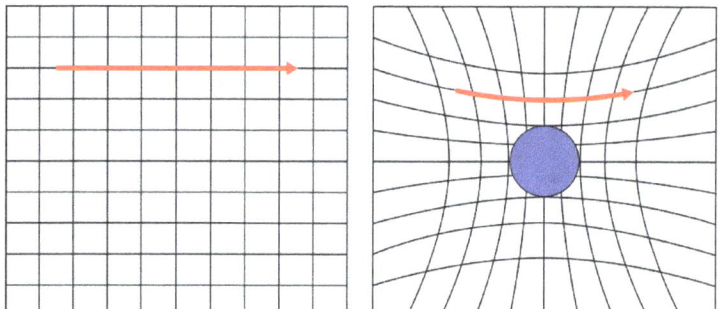

Figure-23. A light beam moving from left to right in empty space (left) and passing Earth (right). If the space is empty, it follows a straight line where its course is bent on the right.

There is one important thing to understand at this point. All the curved lines around the Earth in Figure-21 and Figure-22, are bent by the mass of Earth, they are not straight anymore as in the empty space case. However, in reference to an observer standing just next to the light beam, all those lines, the light beam, everything would look straight. What does it mean? Are those lines bent or straight?

As an observer, if you are standing just next to the light beam in Figure-23, where Earth's gravity is in effect, then you and everything around you are also affected in the same way. The whole space and everything in it are standing on bent lines (i.e., curved Carrier Waves). That's why you don't observe a curved space or curved objects around you. The coordinate lines in Figure-22 and 23 are in effect, they are your references and **anything along/parallel-to these lines would look straight** to you. Just right now, when you are reading these sentences, you already are in Earth's gravitational field in a bent slice of space and any straight object around you stands in a (slightly) curved geometry but they look straight.

This is very important to cross check the Double Existence Model's integrity. The existence is defined as the projection of Material Waves on space-time (Carrier Waves) in DEM. Now we can see, **the physical rules and geometry follow the lines drawn by Carrier Waves** in ST. This confirms DEM's perspective about the way the universe exists and how its mechanics works, because DEM analyzes both the existence and the mechanics in the universe as the interaction of Material Waves with Carrier Waves. Thus, it is expected according to DEM that the Carrier Waves constitute the base of all physical event and the geometry in the universe.

The next item to be discussed would be the reason why objects with mass bend the space-time. In Figure-13, the Ultimate Universe before Big Bang is shown. CES and STK exist, and their touch was defined as the Big Bang in section 4.1.

117

At that moment there was a singularity, which started to gain volume and got bigger after Big Bang, and this was enabled by the effect of Carrier Waves, as explained before. When there is no other factor, they build a straight orthogonal Space.

The Material Waves which create objects with mass in ST, do affect this process and they distort the flat, orthogonal fabric of Space. As seen in Figure-22, the distortion is not random, rather it occurs only in one way by weakening the cosmic pressure effect of Carrier Waves. As you can see, the straight coordinate lines on the left of Figure-22 get closer to each other when the Earth takes place. This is a kind of reverse effect created by the Material Waves. In the case of a black hole, the effect is so strong that it almost creates a singularity in the universe. It is not possible to mathematically express this effect because we don't really know the wave functions of neither Material Waves nor Carrier Waves. However, it is obvious that when the material waves are projected as matter on ST then they cancel out or weaken the effect of the Carrier Waves.

In Figure-19, we have shown two types of Material Waves. The red one (Photon) and the green one (Matter) have an orthogonal relationship, the angle they are projected on ST is different. Just to make an understandable analogy, a sample scenario could be: "the phase of the n-dimensional green wave interacts with Carrier Waves so that it cancels them out where the red waves never interact the Carrier Waves that way because of their angle of projection". Many similar mechanisms can be listed as examples. The main idea is that some material waves interact with Carrier Waves so that they weaken their "cosmic pressure" effect.

118

> Objects with mass bend the Space around them (Gravity). This is Material Wave's effect on Carrier Waves, weakening their "cosmic pressure" attribute. (H15)

Completing the analysis about the interaction of mass with space, we can go to the next step, the mechanics of Gravity. How does the apple fall?

The definition of time in (H9) states, "there is an empty Space for each moment". This structure was shown in Figure-15 and Figure-16. When we apply time in these figures to the gravity models in Figure-22 and 23, then we obtain a more advanced model of the Ultimate Universe, representing both time and gravity.

This presentation is shown in Figure-24. All the Material Waves which constitute the Earth are projected on $Space_{t_0}$ which is the Space of the present time, t_0. Because the Material Waves project an object with mass, they distort the "cosmic pressure" and bend the space. At the same moment, $Space_{t_1}$ is empty because it's not its turn yet. There are no Material Waves projected on it. So, the geometry formed by Carrier Waves are in charge: the empty, flat, orthogonal Space. At the next moment t_1, the Material Waves will be projected to $Space_{t_1}$. They will distort the geometry of it, and it will look like the $Space_{t_0}$ in the Figure-24.

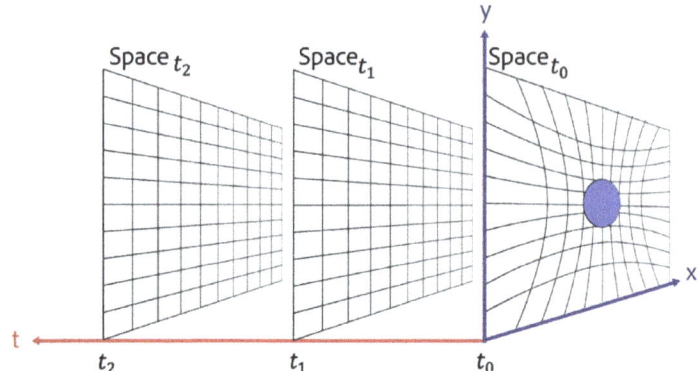

Figure-24. The Ultimate Universe, U^U, including gravity and time. The Earth is projected on ST at the present time t_0, thus, the Space is bent. The space of the next moment $Space_{t_1}$ is empty and its geometry is flat.

So, as all the matter in the universe moves towards future to the direction of empty spaces of next moments and is projected on them, it distorts their flat/orthogonal fabric one by one.

What does it mean for the falling apple? What does happen during its fall? If the effect of mass is as shown in Figures.22, 23 and 24, then how does it cause the gravity effect we observe?

Figure-25 shows the first step of a basic model of the falling apple. The small red circle at line 4 is the apple. On the right picture, we see the Earth in the center, as in the previous figures. You can interpret the vertical numbers, 0 to 4, as coordinates or height of the apple like 4km, as an example.

So, at the first step, on the left, the position of apple's projection on ST is shown at the coordinate line 4. In this left image, the Material Waves are not projected to Space-time yet. The apple is not there. Only its position is shown to be used as reference in the following steps. The Material Waves of the apple are in CES and about to be projected to this moment. They have specific position (let's call it point "A") in CES and projection location of point "A" in ST is coordinate line 4. There is a one-to-one correlation between the positions in CES and ST.

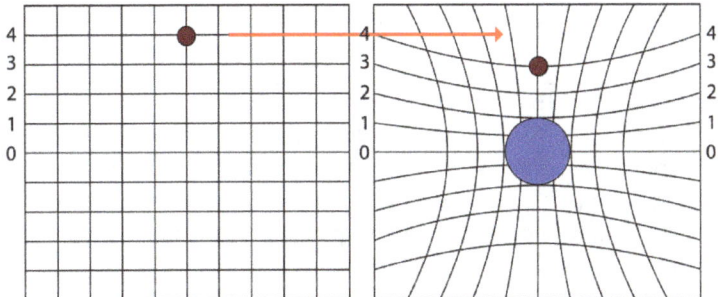

Figure-25. The first moment of the falling apple. On the left, the reference projection location of the apple is shown in empty space. On the right, the universe is projected in that moment. The earth takes place, bends the ST. The apple is still at coordinate line 4 but it is closer to Earth.

On the right image of Figure-25, we see the moment of projection. Material Waves are projected to the empty space of that moment and fabric of the space is bent. Because point "A" in CES is to be projected to the coordinate line 4 in ST, the apple is projected there. On the other hand, the coordinate line 4 is bent and it represents a closer point to Earth, compared to the case of empty space. This results the apple to appear at a lower height in ST even though it stands on coordinate line 4. So, the apple's location is changed in ST which also means its Material Waves are also moved in CES to a new point "B".

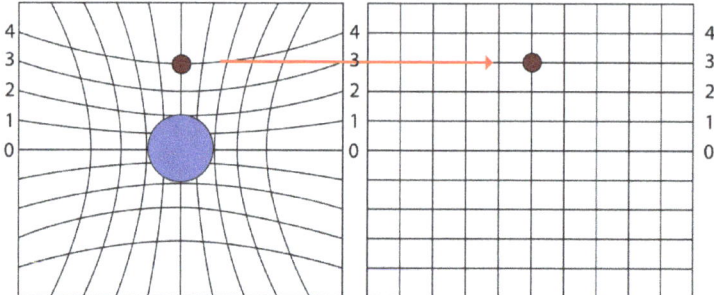

Figure-26. The transition from the first moment to the second moment of the falling apple. The apple is at coordinate line 4 at the first moment in the bent environment. This new location of the apple corresponds to coordinate line 3 in the empty space of the second moment.

Figure-26 shows the transition to the second moment. The apple is at a lower altitude now and its new location corresponds to the level of coordinate line 3 in the empty space of the next moment. So, the apple's Material Waves are at a new point "B" and the apple will be projected to the coordinate line 3 in ST.

The second moment is shown in Figure-27. The apple's Material Waves are projected from point "B" in CES to coordinate line 3 in ST but the apple's height is at a point between 2 and 3.

122

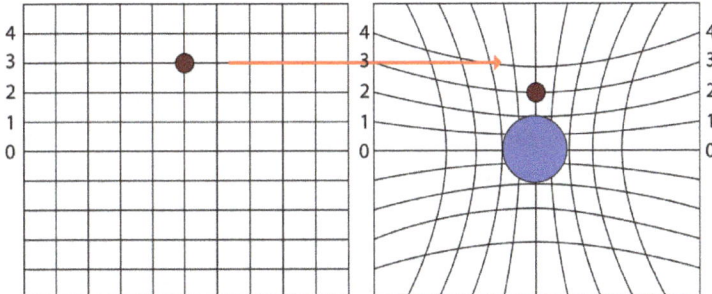

Figure-27. The second moment of the fall. The apple is projected to the coordinate line 3. On the right, where Earth bends the ST, we see that the height of the apple is somewhere between 2 and 3.

This process explains how gravity works according to DEM and is a good example how the mechanics in ST affects (the causal feedback in Figure-11) the complex waves in CES.

4.6- Mass and the Time

Another important phenomenon that is defined by General Relativity is the effect of mass on time. Like moving at very high speeds, if you are close to an object with big mass or in a gravitational field, time gets slower. In fact, this is also valid for objects with small amount of mass, but the effect would be neglectable, almost zero. So, any object slows down the time around it. The earth's gravity, obviously, causes time dilation, too. That's why the earth's core is almost 2.5 years younger than its surface. Time flows faster on earth's surface because its gravity is weaker there.

In section 4.5, we analyzed the effect of mass on space where we found out the effect of the Material Waves on Carrier Waves. Carrier Waves are responsible for the creation and expansion of both space and time. Thus, the weakening effect of the Material Waves which bend the space, are expected to have a similar effect in time dimension, too. They neutralize the expansion in time dimension, so that either the number of projections or their effect on Material Waves are decreased.

As mentioned in section 4.2., time is not a geometric dimension. That's why it may cause misinterpretation to analyze it in coordinate systems and put sequential instances of time next to each other from left to right. Unfortunately, sometimes this is the only way to show and explain the logic. Figures.28 and 29 show space instances of sequential moments in a horizontal raw and the effect of material waves are shown in a geometric methodology but this doesn't mean the effect is meant to be geometric as shown in the figures. This is just a symbolic representation of process and its logic and a helpful way to imagine.

124

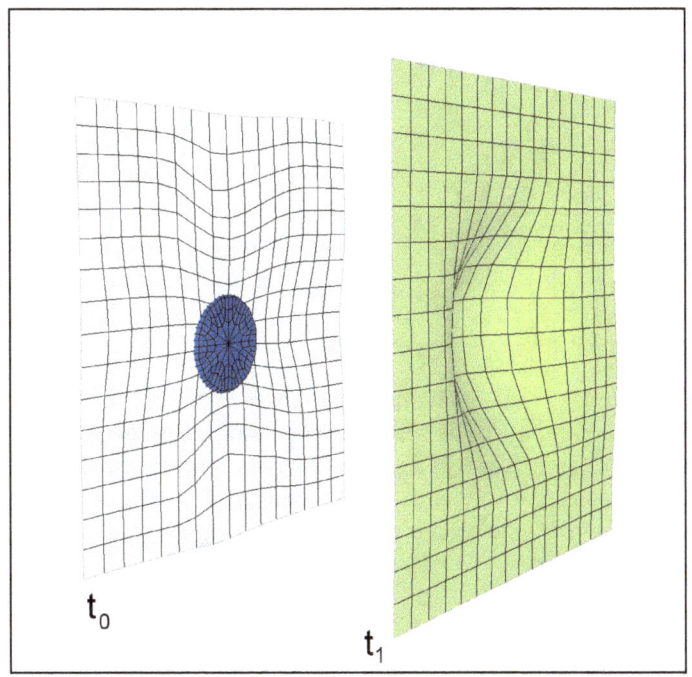

Figure-28. Time dilation effect of mass. The universe is at time point t_0. The Earth is shown symbolically in the center. It distorts the Space around it. The Space of the next moment t_1, is empty and waiting for its turn. Its 3D geometry is flat, but it is bent towards to the Space of the previous moment in time dimension because of Earth's Material Waves.

Both figures represent the same event. As for the case of geometric dimensions, there must be a kind of "distance" (not necessarily used as a geometric term) between space instances where they are spanned along the time dimension. There must be some difference which discriminates a moment with the previous and next ones. Let's call it "time distance".

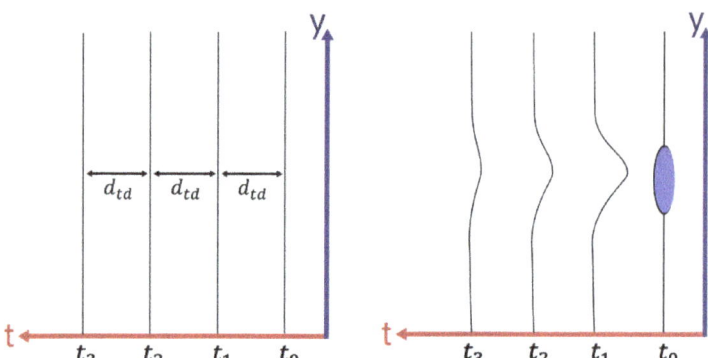

Figure-29. Time dilation effect on multiple instances of Space. On the left, all space instances are empty. They have a fixed "time distance" d_{td} between them. On the right, earth is projected on the current moment. This causes local changes of "time distance" in different parts of the universe.

The cosmic pressure creates the expansion in space-time which is also the root cause of the "time distance" between empty space instances. The "time distance" must be fixed and the same for all the instances. Let's define its fixed value as "d_{td}". It's shown in Figure-29 on the left. The earth is put into the image on the right as source of gravity. This doesn't change the fixed "time distance" d_{td} between space instances but it creates local deflections, as shown on the right. It may even effect more than one space instances of future if they exist.

Neutralization of the "cosmic pressure" of Carrier Waves by Materiel Waves in time dimension occurs this way. How does it cause time dilation?

126

There are two possible scenarios:

1. When there is mass in space, it causes some regions of space instances to come so close that some of them disappear. Thus, the number of new space instances is decreased locally and there are less projections and state updates of Material Waves in that region.

2. The amount of change, the projection/state update causes, depends on the length of "time distance". If it gets shorter as shown in Figure-29, then time gets slower in that region.

Both scenarios can be unified in one model where the local value of "time distance" is set as the main parameter. The first scenario corresponds to the case "time distance=0" and in the second scenario the "time distance" has a positive value which is smaller than its standard amount d_{td}.

<table>
<tr>
<td>

Objects with mass bend the Space-time in time dimension where they cause local change of "time-distance" between space instances of different moments. This results Time Dilation because of gravity.

The amount of dilation depends on the amount of change of "time distance" compared to its standard value d_{td}.

</td>
<td>(H16)</td>
</tr>
</table>

We need to focus on one last thing to finalize all the definitions and discussions about time dilation. CES is an analog medium as we understand which means its existence is continuous, the complex waves exist without any interruptions in CES. The universe is modelled as a discrete time system in DEM, meaning, the universe is technically "sampling" the information from CES like taking screen shots. That's what happens when the Material Waves are projected on empty space instances of new moments.

There is a process going on in between, which causes the change in the universe, resulting the Material Waves' wave functions to have new solutions and states in the next moment. This occurs because of an interaction between Material Waves and Carrier Waves.

This interaction's outcomes are:
- Carrier Waves span the empty Space for the projection of Material Waves,
- When projected, Material waves weaken the cosmic pressure of Carrier Waves,
- This changes the geometry of flat empty space, causing gravity,
- It also changes the "time distance" between empty space instances decrease the effect of "state update process" which causes time dilation.

In (H12), time dilation because of high speed was linked to the number of state updates. In this section, we see there is an additional factor, which is related to the effectiveness of state update depending on the "time distance" between space instances. Unfortunately, it is impossible to build mathematical expressions for these and show how they work in a combined way.

128

However, we can make use of an analogy to describe how the variables of time dilation affect the speed of time. It is like the calculation of the Work when you push an object with a specific weight for a certain amount of distance. The work is the multiplication of the force (which is used to push the object) and the distance. Both factors are proportionally effective. The amount of "time passed" for a wave function, particle or object can be modelled similarly.

1. The effect of temporal speed v_t[22] is like the effect of the force. It represents the temporal power in charge. In Section 4.4, it is explained that v_t determines the number of projections possible. It is also possible that its value determines how big the change per projection will be. So, v_t has two potential effects on time dilation.

2. The effect of time distance d_{td} is like the effect of the spatial distance in the "Work" formula. It determines "how long" the power of the temporal speed is in effect. If it is zero, it means there is no projection regionally in that part of the space. If it is not zero, its value determines the amount of "time passed" per projection. So, d_{td} has two potential effects on time dilation.

---

[22] Figure.21

These two items result a combined model for the two causes (spatial speed and gravity) of time dilation:

Both the temporal speed and time distance may affect the amount of time dilation by
- dictating the next projection to occur or not.
- determining the amount of "time passed" per projection.

5- Micro Cosmos

The ideation of the Double Existence Model was originated from the electron's journey around the nucleus of an atom. This is explained in Section 3.3 and in Figure-9. That's why many phenomena and examples from quantum world are used to make definitions of DEM in Section 3.

In Section 3, there are also explanations about
- the quantum level events shown using Feynman diagrams and how they are modeled in DEM,

- the structure of the generalized wave function $f_{CE}(\vec{x})$ which is the origin of the quantum wave function $\psi_{ST}(\vec{y})$. ($\psi_{ST}(\vec{y})$ is the projection of $f_{CE}(\vec{x})$ passing through the system $Q(\vec{a})$ in Figure-11.),

- the role, the Quantum Mechanics takes place. It is defined as the rule set of Material Waves' entrance to the universe (space-time).

These are some of DEM's fundamental outputs about Quantum Mechanics, explaining the relationship of the probabilistic nature of quantum world with the causality of the macro world and the way the sub-atomic events occur.

This section includes some additional items to discuss about the quantum world to analyze them using the perspective of DEM.

The causal vs. probabilistic universe discussion is generally covered and explained in Section 3. Still, there are items about this concept to discuss in detail which will help to understand the logic better and explain some interesting quantum events which conflict with our common sense.

Let's take an object as an example. This time, the sample object is yourself as a human being. You are made of approximately 100 trillion cells and one cell contains roughly 100 trillion atoms. This means, a human body, your body is made of 10^{28} atoms. If we go to the sub- atomic particles, to quark level, with an estimation of 100 particles per atom, we can roughly say that there are 10^{30} particles in your body.

This quick calculation shows us that you are made of roughly from 10^{30} pieces of Material Waves with 10^{30} complex wave functions in CES environment. They are projected on space-time in the form of you and probably they have a correlated shape in CES, too. The form of your origin in CES is not exactly you, at your size, but it is your original shape in CES, shaped by space-time. Your existence in the universe is a projection, a shadow of it. That's valid for any object around you.

Those 10^{30} complex wave functions are projected on space-time as probabilistic wave functions and act randomly, but they don't randomly split up from your body and go away. Your form in the universe stays solid and stable. This was explained in Section 3 with a mechanism where the positions and all the mechanical attributes of the particles in ST create a feed back to CES, determining the initial/boundary conditions of the complex wave functions in CES environment. So, when these Material Waves are to be projected on to the next moment, they have a probabilistic behavior within their micro world but they are bound to the mechanical attributes of the macro world. A very simple example would be, if an electron is a part of your left eye, it is not projected at your nose at the next moment[23]. It will continue to be a part of your left eye, but it may appear at any random point around the nucleus it is bound to.

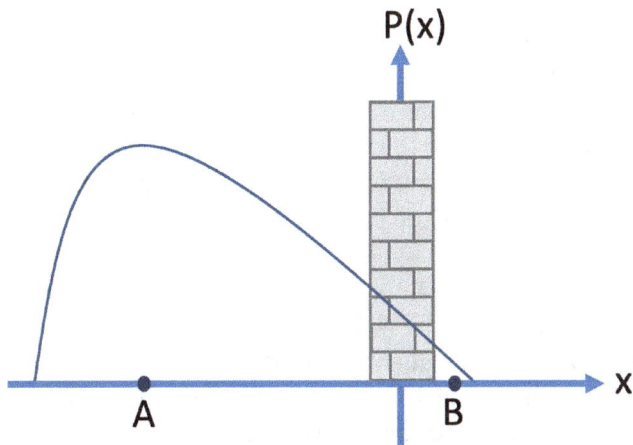

Figure-30. Quantum Tunneling example. The curve represents the probability distribution function of the location of an electron. There is a wall (e.g., it's an energy level which electron cannot pass through) along the vertical axis. The electron is at point A at the beginning. Normally, it cannot pass the energy wall, but it may still appear at point B randomly.

Another interesting example about quantum world is the Quantum Tunneling. Let's take an electron case in consideration. This shown in Figure-30. The electron is at point A. Because of its probability distribution function, it is

---

[23] In fact, theoretically it can appear at your nose if its wave function allows it. But this would be a so low probability that it doesn't happen even once in your lifetime.

normally possible that it may go to point B but there is a (energy) wall in between which the electron can never pass through. However, Quantum Tunnelling tells us that it can still appear at point B and this really happens in real world.

Let's compare it with one real life case. Assume, you sit in your living room, facing to the wall between the living room and the kitchen. So, there is the kitchen behind the wall against you. Start throwing balls to the wall. They all will bounce back from the wall and stay in the living room. No single ball will ever pass the wall and end up in the kitchen.

In summary, quantum tunnelling means that, the electrons can sometimes pass the wall and go to the kitchen. That's one of the points that quantum mechanics conflict with our common senses. How is this case explained by DEM so that we can create a logic and solve the conflict?

As explained in Section 3, the real existence of the complex waves is in CES environment, they are not limited with the physical conditions of the universe or space-time. If the Material Wave of the electron is projected at point A at a specific moment, this is realization of its wave function for that instance of time. We can see in Figure-30 that there is a big probability that the electron can be at point A. We also see there is a small probability that the electron's Material Wave can be projected at point B. So, at the next moment, the electron's material wave function can be realized and projected at point B, without being affected by the wall, because it is NOT in space-time in between and it doesn't have to go through the wall to reach the point B. This case is quite like the example in Figure-9.

One thing which cannot be defined within the context of DEM is the trigger or the mechanism that causes a Material Wave to be projected on ST at any given moment or not. That trigger is probably something beyond our physical perception and understanding. It may be something related to the conditions in CES, or the momentarily states of the Material Waves. These are some logical assumptions but we cannot determine the exact mechanism.

Even though it may be possible to get clues about the trigger or mechanism using some clever experiments in the future. I think so because the cases where we use the term "particle/wave dilemma" are totally referring to this problem.

The double slit experiment is a good example of this. The electrons act like waves in the experiment and establish an interference pattern at the wall behind the double slit. However, if you put a sensor to track their paths, they start to act like particles and the interference pattern disappears. How do they know that we are watching them and change their behavior? That's one of the weirdest things in quantum mechanics.

According to DEM, the electrons' Material Waves are in CES. They may be projected as an energy wave or particle on ST or stay totally in CES environment for some moments. Apparently, electrons are in wave mode along the experiment, normally, that's why they create the interference pattern. Whenever the sensors are in place, the trigger is in charge and the Material Wave is projected as matter on ST. This means that the sensor changes the interaction of the Material Waves and the Carrier Waves and make the Material Waves to be projected as matter on ST.

One other thing we cannot determine within the concept of DEM is the "norm frequency" of the projections in particle level. Let's say there are 1000 empty space instances for every second. Does the time speed one second/second refer to 1000 projections per second? Do all types of Material Waves have the same projection number per second so that they experience one second/second? Maybe, electrons must be projected 500 times and photons 900 times.

If the "norm frequency" is lower than the 1000 projections, it becomes a very interesting scenario. Let's say it's 500 which constitute our journey in time as one second/second. That would mean we exist all the time in CES as Material Waves, obviously, but only half of our existence in CES is projected on ST which we perceive as our whole life.

One of the mysteries about Quantum Theory is standing in its name, the word quantum. "Quantum" is a Latin word, meaning, "a certain amount" (or, how much). The reason why this word is used to name the theory depends on the common property of all particles, or so called "energy packages". For any given type of particle or energy wave, there is always a unit amount. The mass or total energy of such a system is always an integer multiple of that unit value.

Let's take a photon as an example. The photon's energy depends on its frequency. For a given fixed frequency, all the photons in the universe have the same amount of energy. Likewise, every electron in the universe has the same mass. If you build system bringing some electrons or photons together, the total energy/mass will be the integer multiples of their unit amount. In summary, in micro world, everything is quantized and constituted by some packages which are the same size (mass or energy).

136

In fact, there is no attribute we could differentiate one electron with another which means all the electrons are identical. Obviously, this is valid for every particle not just for electrons. Take a Helium atom with two electrons around its nucleus, as an example. If somehow these two electrons swapped their places, we couldn't distinguish them. Or if one of them would vanish suddenly and a new one would place itself magically instead of it, we couldn't recognize that this was a third electron. All of them are like perfect twins or clones of each other.

How is this possible? Why do all the Material Waves of all kinds of particles always are projected identically? The answer to this question is already given in different parts of this book without addressing this specific case. The projection of Material Waves on Carrier Waves, voxels of the universe, the effectiveness of the projection process in time dimension are connected to these phenomena. Let's go over the process step by step:

- In Figure-12, we see a sample Complex Wave Function of an electron in CES environment.

- It passes through the system $Q(\vec{a})$ in Figure-11 which represents the rule set of the projection process.

- The projection will occur on a specific number of pixels (2D example, as shown in Figures-14,15,16,17) which have specific, fixed sizes.

- This leads us to a quantized fabric of space spanned by Carrier Waves.

- Material Waves with always the same set of parameters interact with this quantized space via

$Q(\vec{a})$ which is supposed be also fixed for a given process.

Thus, the same transformation (projection) is applied to the same wave function on the same size of space results always the same result and that's why everything is quantized in the micro world.

Quantum entanglement is the last item to be analyzed in this section. If two particles are entangled, even if they are away from each other at a very long distance (e.g., 1 million light years), there is a connection between them which links their quantum states. If there is a change by Particle-1, its related result is applied to Particle-2. This occurs instantly before any information can be carried between them as if it moves at an unlimited speed.

The brief definition of quantum entanglement, as described above, contains the neglection of two factors in space-time:

- The geometric distance between particles. Both particles change their quantum states as if they stand next to each and are affected by the same event. Let's use a simple real-world analogy. Assume you wash an apple in your kitchen. It will get wet. If you wash two apples at the same time, both will get wet. This is normal. However, a third apple in your neighbor's kitchen will never get wet because of your action in your kitchen. This is exactly what happens between entangled particles. If one of your apples is entangled with your neighbor's apple, it will get wet whenever you wash yours, as if they stand next to each other.

- The limit of speed of light. Beyond the mysterious fact that both particles are affected by same trigger despite the distance between them, the more interesting thing is, it happens simultaneously. This means that the information or the physical effect which modifies the first particle, must reach the second one in no time, moving theoretically beyond the speed of light.

These two items, the geometrical distance and the limit of speed of light are effective for things in space-time, not in CES. Both limitations, which confuse us about the quantum entanglement, are not factors in CES environment according to DEM. The entangled particles' real existence is in CES, and their existence in space-time is just a projection of it. Thus, the entanglement experience we have in ST is caused by some common or correlated components of the particles' complex functions which interact and where:

- their interaction is not limited with the speed of light,

- these complex functions are not geometrically far away from each other as we understand.

in CES environment.

6- The Story of Everything

This section explains the Double Existence Model in plain English, without mathematical expressions and using daily life examples as much as possible[24], just telling the end-to-end story of the universe, The Story of Everything.

Naming the Whole Existence

To find proper answers, first, we need to look at the whole existence, including everything "beyond" the universe and try to define it. We must find a name for the "whole existence". The only thing we know which exists for sure is the universe. That's why we called it the "universe", meaning "everything, all".

However, some other "things" may also exist which we are not aware of. How should we define the whole existence so that we can include every possible existing thing, including our universe? There are endless possible scenarios. There may be other universes, a bigger universe which contains ours in it or totally different structures beyond our imagination.

How can we make such a definition without making up any nonsense rule sets or deductions? Any relation or rule set within such a definition must be avoided to prevent mistakes.

---

[24] The model, scientific ideas and comments are just explained according to the logic of DEM as it is in this section. There is no reference to other scientific data or theories to verify. This is done in previous sections.

We don't know anything about what is beyond the universe, thus we cannot define any physical or geometrical rule.

The set representation is a suitable solution to this problem. Sets are just the lists of their elements and don't define any rule between them. They just represent the existence of their elements and nothing more. So, the whole existence can be represented as a set. We call this set the "Ultimate Universal Set", U^U, in the Double Existence Model. That's the name of the "whole existence".

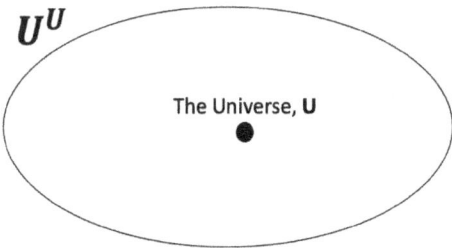

Figure-31. The Ultimate Universal set U^U, as far as we know. The Ultimate Universal Set represents the whole existence. Its only element, we know, is our universe. Thus, the universe is the only element shown in the figure.

Figure-31 shows the most basic version of U^U, where the only known element, the universe is listed as an element. Along this section, as we develop the story of everything, we'll split up the universe into its components, explain how space and time is created and add more elements to this figure. In a way, while telling the story of everything, we'll also draw the detailed version of U^U.

The Double Existence Model

Let's begin with Double Existence Model's main hypothesis. As far as our daily experience is concerned, we live on planet Earth, which is a part of the Solar System, floating in space together with the stars of our galaxy, the Milky Way. So, like every other object around us, we live, stay and exist in the universe which "contains" everything. It doesn't only include our own bodies, the planets, or stars but also all the energy, we observe.

For example, if a light beam leaves a star, travels for 1000 years towards Earth and reaches your eyes, it traverses all the path from that star to your eyes along these 1000 years. It passes each point of space in-between and continues to do this for 1000 years. Along this journey, it is always in the space between that star and you, directed and coming towards you all the time.

In this story, you, the star and the light beam are always in the universe. All of them exist in the space or in the universe and this is our absolute understanding about our existence. We think that we exist where we are. We observe our existence, occurring in the universe with all its components. The electrons around the nucleus of an atom, the energy fields which create the particles, the photons, the cars, the buildings, anything that exists, exists solely and always in the universe.

It's a very natural perception. We live in a 3D space, move in it, design mechanical instruments and make them work. Everything is happening within the 3D space from the beginning until the end. Thus, it is not surprising that our personal perception and all the physical models assume this.

In Figure-31, even I only put the universe as the only element of U^U. What else should exist?

This is where the first and most important chapter of the story of everything moves in. According to the Double Existence Model, none of the matter or energy we observe in the universe or around us, is in the universe. There is a so called "Complex Energy Source", **CES**, which is the external source of all the matter & energy in the universe or space-time. There are so called "Complex Energy Waves" in CES. The complex energy waves are always in CES environment and they are projected on Space Time (**ST**) under specific conditions and create their reflection as matter or energy in ST. Space-time and CES are two completely independent[25] structures which are in interaction.

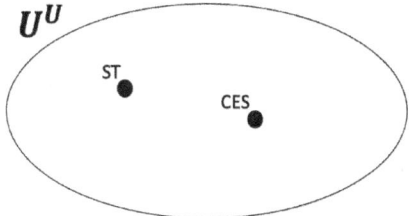

Figure-32. The Ultimate Universal set U^U, showing the Space Time and the Complex Energy Source as its elements. The universe we observe is represented by ST where CES is a secondary element of U^U, generating all the matter and energy in ST.

---

[25] When we explain the Big Bang, we'll see CES has also a role of the creation, and existence of Space Time. The Word "independent" is used here to describe their independent and complementary role while they form our existence.

Figure.32 shows a new version of the Ultimate Universal Set which includes the Complex Energy Source as a second element. Space-time is put into the diagram instead of the universe, however, in this figure it only represents the empty space and time dimensions without any matter and energy in it. In fact, the space-time and the projected energy from CES, constitute the universe as a whole.

Let's clarify the term Complex Energy. There are energy fields which create all the particles we know, like electrons. Every electron has a so called "wave function" which describes its own energy wave. This wave function is basically a mathematical expression. You can use it to calculate the probabilistic behavior of the electron. It defines the characteristics of the electron's energy wave in space-time.

This energy wave is not originated in the universe. In fact, it's a projection of another energy wave in CES, an external energy source, a completely different medium. Thus, the energy wave in CES is expected to have more physical attributes which leads us to the idea, its wave function should have more mathematical components compared to the wave functions we define in our universe.

You can think it like the relation between you and your shadow. Your shadow is a two-dimensional projection of your three-dimensional body which looks like you but doesn't have all the information about you. For example, we cannot see whether your eyes are open or closed if we look at your shadow. That information is lost. Or your shadow is always a tone of gray-to-black, it doesn't reflect the colors of your clothes. So, we could say, you are kind of a more "complex" version of your shadow. Linking this idea with an analogy about ST and CES, we can say, what we see in the universe is

144

your shadow on space-time but the "real you" exists in CES environment.

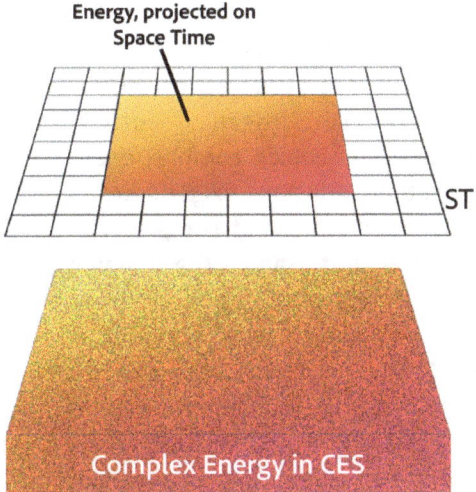

Figure-33. The projection of Complex Energy on Space-time. ST is shown as a 2D-Plane. All the Complex Energy is in CES. A part of it is projected on ST. The square-shaped projection has a correlation to its source, but it is not identical. Complex Energy in CES contains more information and attributes.

Likewise, the energy waves in space-time are a kind of shadow of the energy waves in CES and some information or some attributes of the energy waves are lost in ST. Thus, the energy waves in space-time must be a kind of "simplified" versions of the energy waves in CES. That's why, the energy in CES is called "Complex Energy" and their wave functions are called

"Complex Wave Functions"[26] in the Double Existing Model. It has a more complex structure and energy waves compared to the energy waves in ST.

Figure-33 shows how the projection of complex energy occurs. First, not the whole energy in CES is projected on ST all the time. Each instance of projection includes only a part of the complex energy in CES. The rule set of interaction between ST and CES determine which complex energy waves will be projected each time. It is not a fixed process. Every projection includes a different sub-set of complex waves. The second important thing to understand is the complex waves continue their existence in CES without any interruption and behave according to their complex wave functions and the "physical rule set" in CES. So, they continuously exist in CES and at some moments their existence is reflected on space-time as a trace. A very useful analogy would be an analog camera system. Assume you record a video using an analog camera with a roll of chemical film in it. There are two cars in a parking lot, the first one is red, the second one is yellow. If you shoot the red car first, the frames of the film will record only the red car. The frames represent the Space-time in this example. The parking lot represents CES. The yellow car exists but it is not recorded in your video (not in the frames, not in the space-time.). If you pan the camera to record the yellow car, then it will start to appear in your video (in ST). So, the camera records the yellow car now, not the red one. The projection on film has a different content. Again, the red car is still there in the parking lot (in CES), but only not projected on the chemical film anymore.

---

[26] The word "complex" is not referring to the function's imaginary components.

146

This example explains two things:

- Complex energy waves always exist in CES, continue their existence and their behavior.

- Some variable portion of them is projected on space-time (based on some rules or mechanics) to create our reality of existence.

This main mechanics is the reason why I called the "story of everything", the "Double Existence Model". Our existence is very much like the images on the frames of chemical films, prints of an external reality. The source of all the matter & energy, the raw material of our existence is present in CES, and our reality is its variable projection on space-time, a reduced double of the origin. In other words, there are two copies of ourselves and everything around us, the simplified one in space-time, the original complex one in CES.

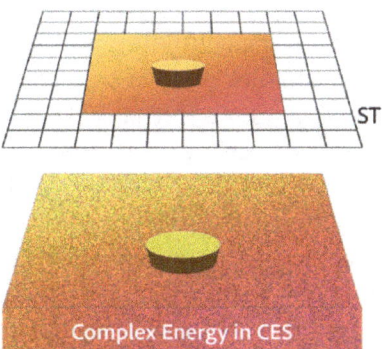

Figure-34. Example of double existence. The cylindric object in ST is a projection of a set of Complex Waves in CES. To represent the correlation, the complex waves are shown in a cylindric shape of a different size.

147

Figure-34 shows a cylindric object in space-time and its Complex Waves in CES. For every object in ST, there is such a set of Complex Waves which create that object's projection on space-time. Both are shown in cylindric shape to represent their correlation but this is only a symbolic presentation because we have no idea about the geometrical properties of CES. The only thing we can say is the complex waves (in CES) which constitute this object (in ST) have a shape (if any) which is projected on ST in the form of a cylinder. In other words, the shadow of those complex waves on ST is a cylinder. They may or may not have a cylindric shape in CES environment.

The projection of complex energy on ST is the first part of the interaction between space-time and the Complex Energy Source. This is the effect of CES on ST.

The second part of the interaction is in the reverse direction. According to DEM, ST creates a feed back to CES where it modifies the states of the complex waves in CES according to the mechanical rules in ST. This happens where the two environments concur, i.e., the projected energy waves and particles. When a complex wave is projected on space-time, then its projection obeys the mechanical rules of it, as long as it is being projected. Any change of the projected energy wave or related particle in ST, affects the source complex energy wave in CES. Thus, we can say the space-time is kind of capturing, modifying, and shaping the complex energy waves. ST is like a tunnel that the complex energy waves must pass, and when they pass it, they are reformed or adjusted.

In summary, the complex energy waves are projected on space-time, create matter and energy and change the fabric of space-time. At the same time, space-time modifies them at

148

the moments they are projected based on the mechanics in the universe.

Let's take the cylinder in Figure.34 as an example. If we move the cylinder to the right in ST, then the complex energy waves (shown in cylindrical shape in the figure) which create it, also move to the right[27].

There is one interesting outcome of DEM about matter and energy. As we know and the most famous physics equation $(E = mc^2)$ suggests that matter and energy can be converted to each other, meaning they are different states of each other, like water and ice. The Double Existence Model suggests that they are two projection forms of the complex energy waves, their mutual origin. This explains their convertibility. They are not just convertible to each other, they are two observable forms of the same thing, the complex energy waves in CES. This is the DEM's explanation about $E = mc^2$ by defining the mutual origin of matter and energy.

Referring Albert Einstein's explanation about special relativity and $E = mc^2$, we see that he also thought, matter and energy were two forms of the "same thing". He said: "It followed from the theory of special relativity that mass and energy are both but different manifestations of the same thing, a somewhat unfamiliar conception for the average mind."

---

[27] The word "right" is a symbolic expression, just to show the correlation of the movements of objects in ST and their complex energy waves in CES. In fact, we have no information about the geometrical environment of CES. So, the direction in CES cannot be defined.

The "same thing" he mentioned, refers to the Complex Energy Waves. Complex Energy Source which contains them, is the "unfamiliar conception" about the existence of an "external" source of all the matter and energy in the universe. They are not "in the universe".

How Do the Universal Mechanics Work?

If you are familiar with physics and its main topics, then you are aware of the discussion about the probabilistic structure of the Quantum Theory. Even though Einstein's work led to the way of Quantum Theory, he refused to accept the outcomes of it. That's why he said the well-known phrase, "God does not play dice". He thought, everything in the universe should be causal.

That's our general experience in our daily lives. Everything is causal and nothing is random. The link between the cause and the effect is solid and always there. If you brake your car, it will always slow down. It will never speed up. If you make multiple experiments under the same conditions and measure the variables, they will be always the same. i.e., if you brake the same way at the same speed, your car will always stop at the same distance.

So, how can such a deterministic universe at macro level could be built on top of a totally random micro world? That is the question.

150

Figure.35 shows the answer of the Double Existence Model to this question and explains how the universe works as a whole system. Let's analyze the figure as a journey of one sample piece of complex energy wave of an electron in your right hand.

1. The complex energy wave is in CES. (Actually, it is always in CES. It never leaves it.)

2. It is projected on space-time. This is the point where the blue right-arrow touches the space-time box in the figure. The wave does not move from CES to ST but as an analogy we could symbolically say, it enters to ST.

3. If it is in, it becomes the dashed-red arrow because it is modified during the projection (or entrance) by the Quantum Level reception system of space-time. It became an observable energy wave or particle in ST environment. This is the phase where the probabilistic behavior occurs. The interaction of the complex waves and the quantum level reception system of ST create the probabilistic wave functions that we face in Quantum Mechanics. The electron of your right hand or its energy wave is projected for that moment of time. It's located at a random place around the atom nucleus it is bound to according to its wave function.

4. The electron is a part of an object, your right hand. This is shown with the box in ST. If you move your right hand, let's say, you raised it, then this electron will also gain some altitude. It is a part of a causal system (your hand) and its location is modified with your movement. Still, its location around the atom nucleus is defined during its projection randomly but it is

always in your right hand and moves with it. This is the causal mechanical effect of ST on objects.

5. You, raising your hand, change the electron's altitude in ST and the feedback mechanism transfers this information to the source complex energy wave in CES which is the origin of the electron. This causes modification of the complex energy wave in CES environment. So, the causality in ST is carried to CES which ensures the continuity of causality.

6. The information of the modified altitude of the electron will be held in the complex energy wave when the complex energy will be projected again at a following moment. So, the electron will have the correct altitude (correct position in ST environment) but can be projected at any random point around the atom nucleus.

In summary, if you raise your hand, all the atoms that constitute your hand will be raised in space-time. This will be reflected to CES and the complex energy waves of your atoms will be "moved" in CES environment, too. Their initial conditions for the next projection will be updated and they will be projected at the correct height. The causality is preserved. Still, they will be randomly projected in micro level.

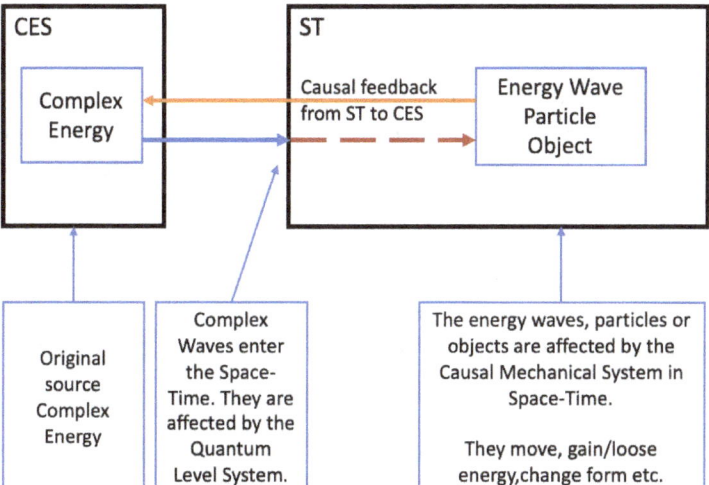

Figure-35. The mechanical flow of the universe. Complex energy waves "enter" the space-time and are projected. This phase is the Quantum-Level process which has a probabilistic feature. The space-time applies its own rules to them which creates the causal mechanics. Its feedback to CES ensures the continuity of causality.

Figure-35 also shows the effective domains of probabilistic Quantum Mechanics and the causal Macro Mechanics. Quantum mechanics is related to the projection phase or the "entrance" of the complex energy waves into the space-time. Macro mechanics are about the events occurring within the space-time. Quantum Mechanics and Macro Mechanics are responsible of two distinct phases of events and are effective one after another. The complex energy wave is first projected on space-time in quantum level, then it is affected by a mechanical system in macro level and updated. Quantum and Macro systems are a kind of "in series" relation. That's why they both can be valid at the same time.

The complete story of the universe should not only explain how it works but also how it started. The Double Existence Model introduces a set of hypotheses which claim the existence of an external source of energy and matter. As an analogy, we could say, this external source, CES, kind of "touches" the space-time. By doing this, it creates projections of its complex energy waves on Space-time and generates all the matter and energy in the universe. How could this process have begun?

Think of a moment of empty space. There is nothing in it. It's completely empty and dark. Then, suddenly, CES approaches the space and the first projection occurs. There is some matter and energy in the space for the first time. Of course, there wouldn't be any stars, planets or galaxies at such a moment because nothing would be shaped yet. It would probably be a bunch of random energy waves. Actually, Big Bang is a similar event. The universe suddenly encounters an energy flow, at extremely high pressure and temperature. More and more energy comes in, expands, cools down, forms matter. So, we could say, Big Bang was something like this but one thing is missing in this story. There was no space at the moment of Big Bang. It was created simultaneously with it. We don't know anything about the history of CES. It may have been existing forever or not. It's logical to assume, it existed "before"[28] Big

---

[28] Time is something related to our universe. Thus, the phrase "before Big Bang" has no scientific definition or meaning. However, it is used to be able to describe the event so that the reader can have an understanding about the order of events as our brain likes to perceive.

Bang because we observe the entrance of energy into the universe during Big Bang.

What about the space itself. We know, the space started to exist and expand with Big Bang and it is still expanding. This led us to the idea that, the space was a singularity (had no volume, like an infinite small point) during of Big Bang. So, CES couldn't have "touched" the Space and started the existence in it.

Then, there had to be something else which interacted with CES and caused the Big Bang. It is not necessarily a floating object which is collided with CES and exploded. It can be a "thing", an event, even some phenomena in CES environment. We can't know that. What we can define for sure is that there was a "trigger", a root cause which generated the space-time when it interacted with CES for the first time. This trigger is called "Space Time Kernel" in DEM.

Figure.36 shows the Ultimate universe, U^U, just "before" the Big Bang. CES and STK are its known elements. For an unknown reason and via an unknown mechanism they get into interaction and caused Big Bang. The space started to expand beginning with zero volume. As long as it expanded, more and more energy waves came into it. This process built the universe, as we know it. So, we can briefly define the Big Bang as the first "touch" of the Space Time Kernel and the Complex Energy Source.

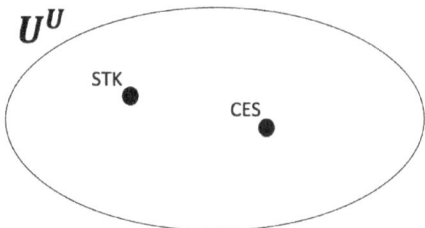

Figure-36. The U^U just "before" the Big Bang. STK and CES interacted for the first time and caused the Big Bang.

What caused the Space to exist and expand at the moment of Big Bang and later? It's a very important question because its answer will help us to understand the fabric of space and thus explain the reason why the universe works as it does. The scientific studies about Big Bang show us that all the affecting factor was the incoming energy. It not only created the matter in the universe but also created some kind of cosmic pressure which caused the expansion of the space. On the other hand, we know the Dark Energy is the root cause of the continuous expansion of the universe.

Briefly, normal matter as we know and its mass creates a gravitational effect which pulls everything together and the dark energy shows a kind of reverse effect. If there was no dark energy in the universe, the gravitational effect of all the galaxies would cause the universe shrink to zero volume (singularity) again, eventually. As we know, dark matter and dark energy cannot be directly observed as normal matter and energy. That's why they are called "dark". However, we know they are in charge in the cosmic system.

Why are they not directly observable? Why can I not see, hold, or weigh a piece of dark matter? Why is the dark energy different? Why does it generate a kind of cosmic pressure and expands the Space?

The answer lies under the different types of complex energy waves in CES, according to the Double Existence Model. As mentioned before, the energy waves in CES are called Complex Energy Waves because they have more attributes than the energy waves we observe. There are probably many different types of them, interacting with ST in different ways.

There are two types of complex energy waves defined in the Double Existing Model:
1. **Material Waves:** These complex waves are projected as matter and energy. They are the building blocks of ourselves, the stars, and the galaxies.
2. **Carrier Waves**: These complex waves span the space itself. They constitute the fabric of the space. Their existence is directly the existence of space.

Carrier Waves are called after the carrier waves in telecommunication and broadcasting. Real audio and video signals are carried by using high frequency sine waves in these technologies. For example, if you tune an FM radio channel at 100MHz, it means the music you are listening to is added to a 100MHz sine wave and broadcasted via an antenna. Your radio receiver gets this 100MHz signal, extracts the music of it and plays for you. The 100MHz sine wave is called a carrier signal because it literally carries the real information to you. A predetermined mathematical formula is used to load the information to the carrier signal. The reverse formula is used to extract it. This mathematical rule set is called modulation. The simplest and best visual example, the Amplitude

Modulation (AM) is shown in Figure.37. The sine wave at the top represents the music info, we want to broadcast. The high frequency carrier wave is on the second line. The third line shows how the modulation combines two signals. The graph on line four shows clearly that the carrier signal's shape (mathematically, it's called the envelope.) is the same as the input information, the song. As an analogy, we can say, the song exists on the carrier signal. Assume, it is not a song, but your own live video. Then the carrier wave would be hosting you and be shaped by you.

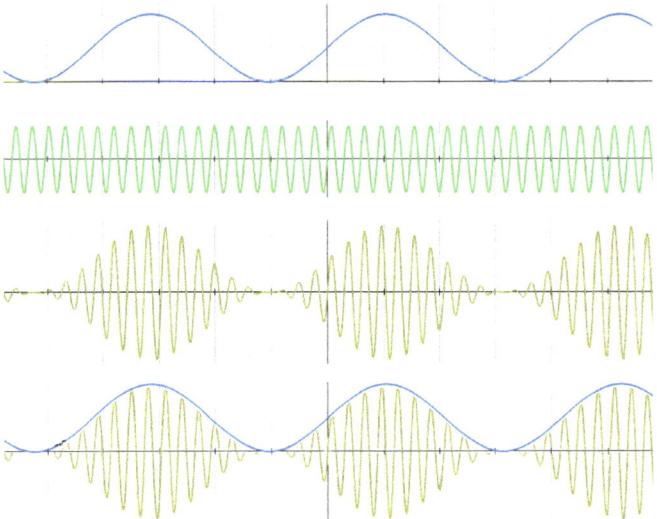

Figure-37. The Amplitude Modulation. The carrier wave's shape is modified by the input signal, the information to be carried. The modulated carrier signal looks like the input signal.

That's the reason why the complex waves which constitute the space are named after the carrier signals in telecommunications. Carrier signals constitute the medium which contains and carries the audiovisual information and carrier waves defined by DEM constitute the fabric of the space and host all the matter and energy. It's a very similar mechanism.

Combining all the information given in this sub-chapter, we can summarize outcomes of the Double Existence Model as follows:

- Carrier Waves have different attributes than the Material Waves and their interaction with space is different.
- Dark energy is the projection of Carrier Waves.
- During the Big Bang, carrier waves started to build and expand the space. Material waves started to fill that space with energy and matter.
- Carrier waves constitute the space and Material Waves are projected on space, i.e., our existence originates from the relation of these two types of complex energy waves. Material Waves create our material on top of the space fabric which is created by Carrier Waves.

Figure.38 and Figure.39 show how the fabric of Space is created by the Carrier Waves. Figure.38 is a one-dimensional example. Assume that the Carrier Waves is just a simple sine wave with a fixed frequency as shown in the figure. The energy (and some other unknown attributes of it) creates a kind of cosmic pressure and spans/creates a certain amount of "place". In the figure, it is shown as each full period of sine wave which corresponds one unit length (one side of the squares in the figure), for the sake of simplicity. So, each single period of sine wave creates one unit length of space. Figure.39 shows it for the 2D case. Where each 2D piece of sine wave creates a unit-area of space.

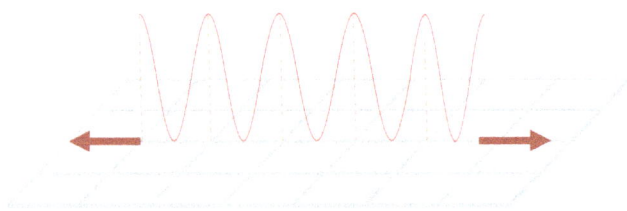

Figure-38. The 1D carrier wave example. Each full period of sine wave creates space of 1 unit length.

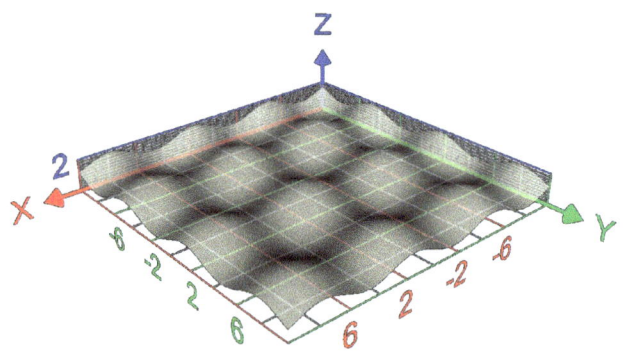

Figure-39. The 2D carrier wave example. 2D sine waves create 2D planes of Space. The coordinate system lines show the correlation of the carrier waves with amount of the created space.

We can use the "pixel" analogy for convenience to go further in the discussion. As you know, the word pixel (short for "picture element") is commonly used in digital photography. It's the number of "points" in a digital picture or camera sensor. The more pixels there are in an image, the more detailed it is, because there are more "points" to show the details. Pixel is a two-dimensional term. The 3D version of pixel is called voxel ("vo" for volume, instead of picture). Using this terminology, we can say, the squares drawn by the coordinate lines in Figure.39 are the pixels of the space in 2D domain. If we convert the concept to the real 3D world, the carrier waves span the 3D space by crating the 3D unit voxels of it.

This is very important. According to DEM, the 3D space is a big set of 3D unit voxels created by Carrier Waves. The size of these voxels should be the minimum possible volume in the universe and the length of one side of these cubes should be the minimum possible length. That leads us to the Planck Length, the minimum possible distance in the universe.

Planck Length's background is not the only outcome of the Carrier Wave analysis in DEM. It also tells us the space is discrete, not continuous. This means, there is no one big space as a whole. Rather, it is a big set of very small independent space pieces.

In other words, the space itself is quantized. That's the answer to another big question in Physics. Why is everything quantized in micro world? Why does every electron perfectly have the same mass or why do all the photons (at the same frequency) have the exact same amount of energy? The answer is in the projection process. When a specific type of material wave (let's say of an electron) is projected on space, its projection must be on a certain amount of space, i.e., certain number of voxels (let's say 2). The input is the electron's material energy wave with specific and fixed attributes, and it is always applied to 2 voxels with the same process. So, the outcomes should be always identical which leads us to a quantized micro universe.

As promised, along with new hypotheses and comments in this chapter, we'll draw the Ultimate universe in more detail. Because the book has only 2D distribution medium, we will continue with 2D Space graphics instead of 3D.

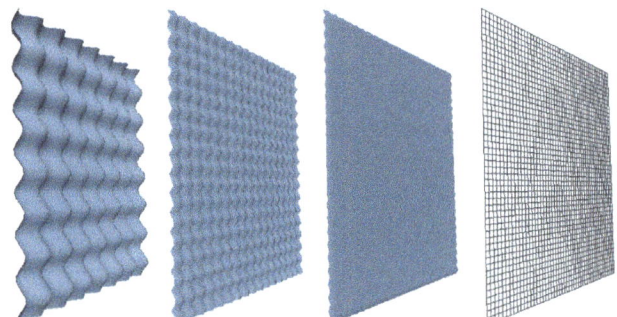

Figure-40. The 2D Space planes, spanned by carrier waves. The first three planes created by carrier waves of different frequencies. The most right one is the quantized presentation of them which also helps to see how the coordinate system is formed in empty Space.

162

Figure.40 is a redesigned version of Figure.39. The sample 2D space planes are positioned vertically. The first three samples show the spaceplanes in the form of their carrier waves. The frequency of the carrier waves is different and it increases from left to right. The aim is to visually demonstrate that higher frequency carrier waves constitute a higher resolution space. This is another important point. The frequency of the carrier waves must be very high to build our universe as it is. Their frequency must be high enough to be able to contain very small things and sub-atomic particles. Using the pixel analogy, you can think, you need more pixels to see more details. To have more pixels (voxels in 3D Space), we need higher frequency signals as shown in Figure.38.

The rightmost 2D spaceplane in Figure.40 shows the pixels and/or the coordinate lines of the 2D Space. It is the quantized version of the other three samples. This is the most useful presentation of 2D Space because it shows both the quantized structure and the coordinate lines at once. Thus, we'll use this version to represent the space in the Ultimate universe in the following illustrations.

Figure.41 shows such a presentation of the Ultimate Universal Set U^U. On the left, one moment of Space-time is shown. Because it is a frozen moment, there is no time dimension shown in this figure. So, this 2D plane represents the 2D space only. Some Material Waves are projected on space which is spanned by the Carrier Waves that are shown as black lined squares on the left.

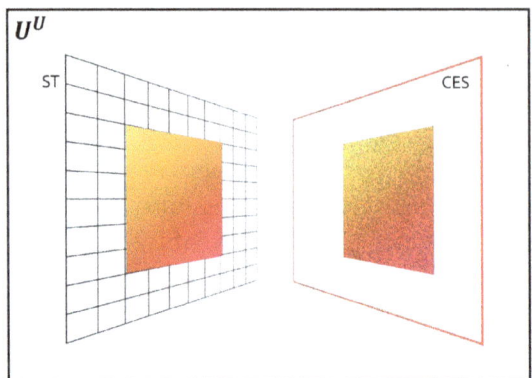

Figure-41. The Ultimate universe, presented with 2D quantized space and 2D CES illustration. The fabric of the space, created by the Carrier Waves is represented with the black-lined squares as a 2D plane. The projection of the material waves in CES is shown on the left.

Time

Time is about change. If there was no change, we couldn't perceive time. Figure.41 shows a frozen moment of time. It's one projection of Material Waves on empty space. This is a snapshot without any change in it. To implement a change in the universe, a new projection of Material Waves on space-time must occur. That will create change and a new moment in space-time. The change may be caused by two things:

1. Something may have changed about Material Waves in CES environment. So, their next projection will be different than the current one.

2. The projection process creates new outputs for the same Material Waves.

164

Both cases are shown in Figure.42. The first projection occurs at the first moment, called t_0. The second projection occurs at the second moment, called t_1.

1. There is change in CES domain at the second moment in bottom right corner which is reflected as a change in ST.

2. There is no change in CES domain in upper left corner, but it is projected differently on ST at the second moment.

$$t_0$$

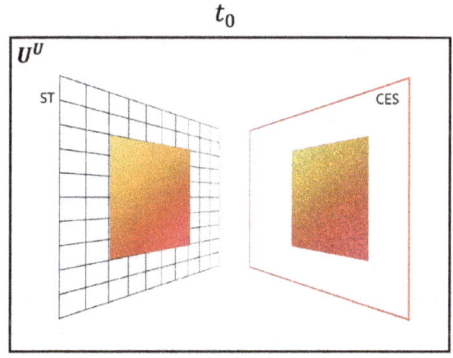

$$t_1 = t_0 + dt$$

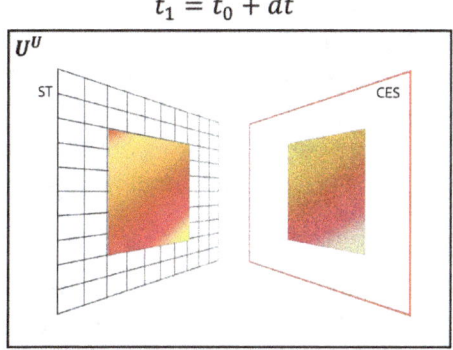

Figure-42. Two moments of projection and the change in Space.

In both cases, the next moment is set by new states of particles and objects in the space as the output of the projection. So, each moment in time is related to the change of the states of each projected Material Wave and its projection on space. This is a kind of state machine.

We can make use of two analogies. The first one is a digital circuit with a digital clock. Time and the flow in the circuit is triggered by the digital clock. Each time the clock creates a pulse in the circuit, each circuit component makes one step of its operation and changes something. This is what you know as the processer frequency of your computer. If it is a 2GHz processor, the digital clock creates two billion pulses to trigger the processor components each second. They all are synchronized, so, all the components change their states two billion times each second, meaning they have two billion different states each second.

The second example is an analog movie camera, recording a video. Let's consider one frame of the chemical film in the camera. This frame represents the "ST" labeled 2D spaceplane in Figure.41. The world outside of the camera represents the Complex Energy Source, CES. When the camera starts recording, it will use this first frame to record the first moment of the scene. It will record the states of all the objects in the camera's field of view. The second frame of the film roll will capture the next moment. Any moving object will be displaced in the second frame where still standing objects remain the same way.

Both examples show the relation of time and change of states using similar flows like the projection of the Material Waves on space-time.

That's where Double Existence Model's definition of time comes from:

"There are empty Spaces for every moment in the universe. Time is the creation of next reality state by complex energy waves being projected on those empty spaces, sequentially."

Like the frames in the film roll, there is an empty space for each moment of time. New moments are projected on those empty spaces and these are created by Carrier Waves as the 4th. Dimension. Carrier Waves didn't only span one big space for us to live in. They create a separate new big empty space for each moment of time.

Figure.43 shows the space-time as a whole structure according to the definition of time[29]. There are multiple copies of empty spaces which is shown in Figure.41. Each one is used for the projection of the Material Waves of one moment. Future empty spaces are waiting for their turn. Each one is used only once.

That's a very huge claim if you try to imagine it. The universe, or the whole Space is so big, and it is continuously expanding. Double Existing Model suggests that there are many[30] other empty copies of this gigantic structure. As human beings, we

---

[29] Because we can only draw illustrations on 2D papers and can only use geometric positioning, the multiple empty spaces in Figure.43 are positioned in a horizontal line from right to left. In fact, time is not such a geometric dimension and empty spaces are not lined up like that. This the only possible presentation on the paper which symbolically shows the idea.

[30] A discussion about "how many" is in next pages.

try to understand the existence of such a big universe which is something magical and very difficult comprehend. On top of that, DEM tells us that there are many of them. Please note that this not something about parallel universes. It's the existence of our unique universe along time dimension.

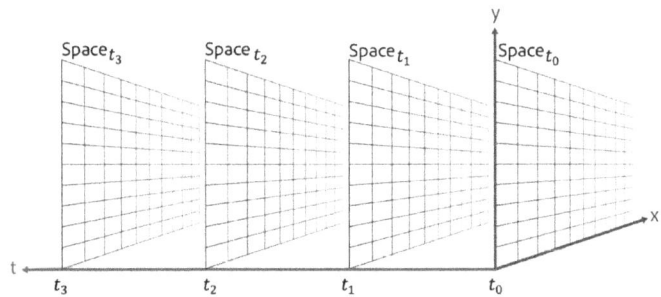

Figure-43. Space-time including both space and time dimensions. Sample 2D spaceplanes represent the spatial space coordinates. Time coordinate is not a similar geometric dimension. It is made of multiple copies of empty spaces. $Space_{t_0}$ is the empty Space for the current moment (present), $Space_{t_1}$ is the empty Space for the next moment. Time flows to the future from right to the left.

A different perspective could be useful to understand this notion. We don't live in the same instance of the space during our lives. On the contrary, each moment is projected on a new, unused, empty instance of space. This is just like every moment of a scene being printed on an empty frame of film roll. If you watch a movie, each frame is projected on the screen and you have the feeling of a continuous event going on. In reality, each moment of the movie is printed on a different frame. It's the same for our universe. Even while you are reading this sentence, you have been changing the space you are in, at least millions of times.

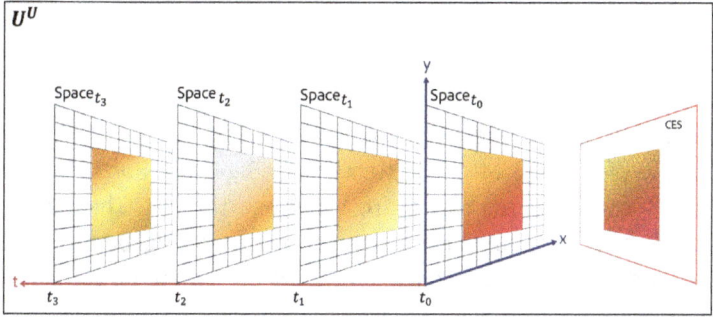

Figure-44. The Ultimate Universal Set, including CES, Space and Time. The Material Waves are projected on empty spaces sequentially and create the reality of the next moment. There is an empty space for each moment, named as t_0, t_1, t_2 and t_3. In this figure, the projection is done for these 4 moments of time.

There is another important thing about the definition of time. The definition describes a discrete time mechanism. Time is not continuous. The change occurs when the Material Waves get into interaction with the next moment's empty space and the projection is done. We can symbolically think that CES is an amount of energy flying through the empty Spaces in Figure.43 from right to left. The projections occur whenever it hits the next empty space and the next moment is created. Again, this must happen very fast at a high frequency but it is not a continuous change. As in the examples of computer processor or movie film, the state changes are discrete changes.

Compiling all the information up to this point, we can draw a new and more detailed version of our total existence, the Ultimate Universal Set, U^U. It's given in Figure.44.

The Amount of available Future

Figure.45 shows the Big Bang and the Space-time. This is a very typical illustration. It starts with Big Bang on the right. Time flows towards future from right to the left. The size of the Space is a singularity at the moment of Big Bang but it expands in time and particles form objects, planets, stars and galaxies.

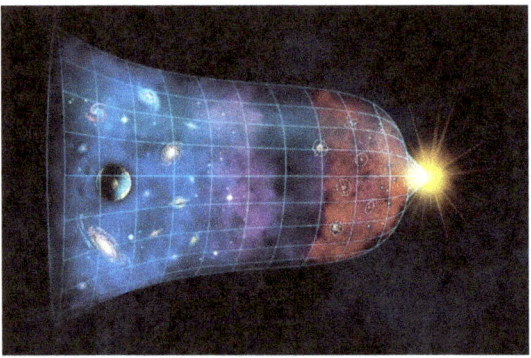

Figure-45. Typical illustration of Big Bang and Space-time. The moment of Big Bang is shown with the explosion on the right. Time flows forward from right to left. Space is shown in 2D and from right to left, it expands in time.

170

Let's compare the Space-time models shown in Figure.44 and Figure.45. There are some important differences between them. They describe two different types of structures of our universe. This is analyzed in Table.1.

Universe in Figure.44	Universe in Figure.45
CES exists as the source of matter and energy	All matter and energy are located within the 3D Space
Space is made of voxels and quantized	Space is continuous
Time is discrete and quantized	Time is continuous
There are multiple spaces to create each moment of time	There is only one space, and it is somehow connected with a continuous time dimension

Table-1. Comparison of universe models in Figure.44 and Figure.45

Despite all these differences, both figures have one thing in common. Because they must fit the size of this book, they have limited sizes. That's why the space is drawn with some limited height or area which is placed vertically. Time is symbolized with limited number of spaceplanes in Figure.44 and with a limited length of the tube-shaped space-time figure in Figure.45.

So, they show some symbolic amount of Space and a limited amount of time. Please, answer the following question if you have seen a similar illustration of Space-time like in Figure.45 before. Have you ever thought how long the length of this tube in time dimension should be? These illustrations always show the time up to now. What about the future?

Have you ever thought how much future do we have?

We have some understanding about the size of the universe, or the space itself. Astronomic studies show that its radius is 46 billion light years. So, we know how much space is there if we start an inter galactical journey. In a simpler way, we know the Sun is 150 million kilometers away from the Earth. If I could walk from Earth to Sun, I would be able to go all these millions of kilometers step by step and literally see that the space in between really exists. I can also see the stars, light years away and I know there is space between me and those stars. We can physically see, observe and measure the amount of Space, created after Big Bang. We can even measure how it expands. There is no doubt about the existence and the amount of the space which began existing and expanding with Big Bang.

What about the time? Time also started to exist with Big Bang. Logically, the time dimension should also be something expanding along with spatial dimensions. So, like the space with a certain radius, the time dimension should or could also have a certain amount of size. Or shouldn't it?

That refers to the length of tube-shaped Space-time in Figure.45. More importantly, it's the number of empty space instances or spaceplanes in Figure.44, according to the Double Existing Model.

This is very important and very interesting. First, if it has a finite size like the space has, then it means we have finite amount of future ahead of us. It may be a very long future as well as a very short one. If that's the case, our existence may end any moment and we would never be aware of it.

The definition of time in the Double Existence model offers a capability to deeply analyze this situation. Figure.46 shows the space-time with empty space instances for past, present and future. There are different scenarios based on the number of each type of them.

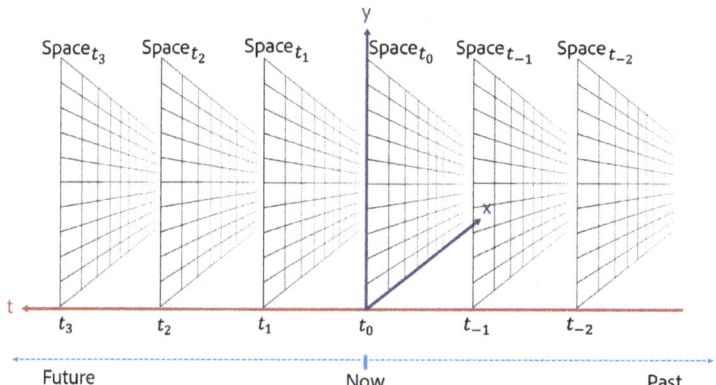

Figure-46. Space-time including the past, present, and future space instances.

The amount of possible future depends on the number of empty future Spaces available. The amount of history which exists, if any, depends on the number of past Spaces which still exist.

Let's analyze some interesting scenarios[31]:

1. The past has occurred. So, all the past spaces existed for sure. However, they may have vanished immediately or after some process. This leads to the conclusion that some or all historic records on past spaces may have gone forever. To make it more clearly, we can say:

    a. Not all the past spaces which begin to exist with Big Bang, may still exist today.

    b. If at least some of them exist, history of one second, one day or one million years may be available. We cannot be sure how many.

2. The present is what we live and observe momentarily. So, it exists for sure along with our existence. This is the only space which exists for sure and there can be only one present space. The only discussion about the present space which is $Space_{t_0}$ in Figure.46 would be about its solitude:

    a. Figure.46 shows the past, present and future spaces all together at once, as if they existed simultaneously. This is a possible scenario but it is also possible that the present space is the only existing space.

    b. If the past spaces vanish as discussed in item 1 and if the future Spaces are created "just in time", then the only existing space is the present one.

---

[31] Full analysis of all possible scenarios is given in Chapter 4.3.

c. That would mean there is no past available and no future ready to live in the next moment.

3. We can never be sure about the existence and the numbers of future spaces.

   a. There may be some future spaces available and waiting to host us. They may guarantee a long or short future (may be only 1 second or billions of years)

   b. As discussed in item 2, there may be none available where each new moment is created instantly.

   c. In both cases, we can talk about the "production" of future empty spaces by Carrier Waves. They started to create them during Big Bang and have continued to create new ones for the last 15 billion years, as far as we know. As carrier waves build and expand the space at a certain expansion rate, they also expand the time dimension towards future at a certain "production speed".

   d. Then the question would be about the amount of the "production speed". As an example, let's assume we consume 100 empty spaces each second. If the Carrier Waves produce less than 100 empty spaces each second, then at one moment, we will reach to the end of time. We will have consumed all the time we had and our existence would end or maybe frozen.

e. For the last 15 billion years, the Carrier Waves have created new moments of time (new empty spaces). Thus, we shouldn't be much pessimistic about this. However, we can never be sure that the process will continue and it will continue fast enough. There is an end of the future we don't know how far away.

The Unique Reality of the Universe

There is one more thing to complete the story of everything and to understand the mechanics and reality of our existence. It's the last thing about the time dimension. Along the book, I repeatedly underlined that the time dimension is not a geometric dimension like the 3D spatial dimensions. I also explained that time is the change of the states of energy waves during their projection on space. Defined like that, Space and Time seem quite unrelated, independent but we define both as dimensions and there is one thing which defines a big and mysterious correlation between them: Time Dilation.

The speed of time is not constant and it gets slower if we move faster in space. That's one of the most famous outputs of the special relativity. General relativity adds another variable into the game, the mass. Object with mass also slow down time. These are two known factors which affect the speed of time[32].

---

[32] The explanation of the Double Existence Model about how the spatial speed and mass affect the speed of time is explained in chapters 4.4 and 4.6. This section only makes an analysis based on DEM's hypotheses.

Science Fiction used this notion to create creative stories. The idea that time is a dimension, made people think, we could travel along it. I think, time travel is one of the most amazing ideas in human history. I can't imagine anything more exciting than time travel.

However, because time is not like a geometric dimension, we cannot easily travel along time. On the other hand, the effect of time dilation is a reality which is observed in space missions and e.g., GPS systems use the time dilation calculations to give your position correctly on earth.

Let's analyze how the time is in charge as a dimension. To make a comparison with spatial dimension, we need to recall some basics of the spatial dimensions. Figure.47. shows a very familiar event. The red point at the center is a car. It makes its way to the North and travels 100km. When it completes this action, it is 100km away from its starting point. Then, it starts to move back to South and travels 100km again. At the end, it is at its starting point again.

Another fundamental outcome of movement in spatial domain is shown in Figure.48. There are two cars shown with red and green dots. At the beginning, they are at the same point, standing next to each other. They start to move simultaneously into the same direction. The speed of the red dot is 60 kilometers per hour and the speed of the green dot is 100 kilometers per hour. After one hour, the red dot is 60km away from the starting point and the green dot is 100km away.

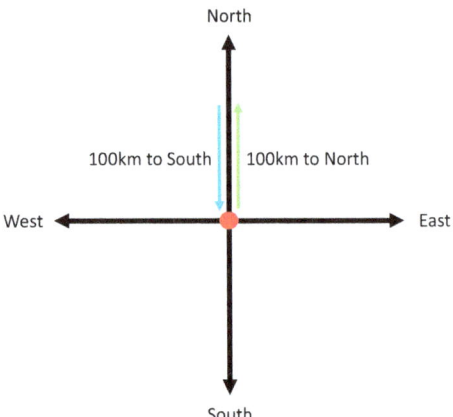

Figure-47. Movement in spatial coordinates. If a car, shown with a red point at the center moves to North and travels 100km first, and makes the exact opposite movement then, it comes to its initial point.

At the beginning

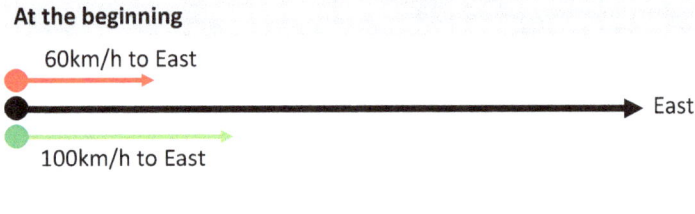

1 hour later

Figure-48. Movement in spatial dimensions takes objects to different positions in spatial coordinates. Objects can change their positions and stay there. If two objects at the same point move differently, they end up at different points in Space.

The important point here is (which will help to understand the difference with time dimension) they stand 40km away from each other. They are not at the same point in space, or on Earth. The drivers of these two cars cannot see each other, cannot talk to each other. They can't have any direct physical contact. Their movement in spatial domain **separated** them.

Time dilation doesn't work like that. If two objects move in different temporal speeds, they are not separated in time dimension like the cars in Figure-48. When they come together physically in spatial coordinates, they are at the same moment in time. Let's recall the famous twin example, explained in Section 4.4. One of the twins is an astronaut, goes to space-trip at a very high speed and experiences time dilation. He leaves the Earth in year 2000 at age 30. He comes back 2050, his twin brother is 80 years old. Because of time dilation, only 10 years passes for him, so he is only 40 years old.

The important thing is this. They moved in time dimension at different speeds and different amount of time passed for them. The astronaut moved only 10 years forward but his twin moved 50 years. If it was like the case of cars, the astronaut should be in year 2010 and his twin should be in year 2050, but this is not the case. They both are in 2050.

The bottom-line of this is, the "time passed" for us is not about how much we moved along time dimension, it is about "**how much we are affected by time**". This is also in line with the definition of time in DEM, change of the states with each projection.

So, think all the objects around you, the planets, the stars, the galaxies. They move at different speeds, under the effect of different gravitational forces. Some of them are so fast, that time flows them a thousand times slower than it does for us on Earth. It doesn't matter, whatever their temporal speed is, where they are in the universe, they are all at the same moment of time, the presence, our presence.

Combining all the components in this chapter, the Ultimate Universe, the Complex Energy Source, definition of time, empty space instance of new moments, the mechanics of time as dimension, we have a conclusion.

CES is the source of all the matter and energy in the universe. The Material Waves in CES contain the information of each particle and energy wave. They are the source and determine the initial conditions of the next moment. Carrier waves constitute the space-time. They built a medium to interact with material waves and kind of "display" the result along each projection. The information about our reality is stored and processed in CES and projected on the empty space instance of "now".

All these hypotheses and deductions describe a universe with a unified reality. Almost nothing in the universe is the same age because of time dilation they were affected by. Though they have been existing since Big Bang. They are part of the same unique reality which is the "present" or "now".

And "now" is the only real moment in the universe.

This completes the conclusion of DEM about the reality and uniqueness of our existence:

- The presence we experience is the real presence and the only live moment in the universe.

- The future does not exist in CES yet. No information is already created about the future and nothing is accessible about it. Time travel into the future is impossible.

- The past has disappeared in CES and it cannot be experienced again or changed. It can be observed only if the previous space instances still exist in U^U. Time travel into the past is impossible.

7- Conclusion and Comments

All the physical theories and models suggest that we are fundamentally made of energy waves, we are energy waves. What we call as matter is a way of appearance of those energy waves. Of course, that appearance has properties of its own, like mass. So, it is not the pure or the original form of energy, rather somehow modified and positioned in space-time with some physical rules and restrictions.

The Double Existence Model tells nothing different about it. We are made of energy waves. The point is they are not in this universe. Complex Energy Source, CES is the fountain which enlightens the space-time with its energy waves. Carrier Waves constitute the quantized fabric of space-time and they host the Material Waves which create the particles and energy we observe in the universe.

Every time the Material Waves are projected on ST, their wave functions create results where quantum level probabilistic character of sub-atomic world occurs. When this happens, Material Waves are trapped in space-time and they obey its mechanical rules which enables causality.

Each projection process is a new moment of time, happening on a brand-new empty space instance. The information of our reality is stored in CES and displayed on ST. That leads us to the fact that, there is no past or future available to go to. As stated in Section.4, there may be a chance to only observe the past.

In this model, there are no parallel universes. We have one CES with our Material Waves which carry the information of our state of presence. There is one universe and only one and real present.

This is my first version of the Double Existence Model. DEM is basically a concept about the universe which is based on the axiom that the matter and energy is not in it. All the other outcomes are deductions of analyzes based on this axiom. Thus, we should except that, some alternative combinations of possible scenarios are equally possible. That's the reason why I called this book a "brain storming session".

For example, it may be possible that there are parallel universes. The Material Waves may have such attributes that they have more than one projection and can store more than one state. In this case, they would be able to create more than one reality which would mean, things like parallel universes and time travel are possible. I think it's a long shot. On the other hand, there are even more alternatives if we go that way. Maybe there are other CES structures than ours. They may have created their own universes. Why not?

I think, such scenarios could offer science fiction a new vision about the definition of reality and its manipulation. As a science fiction fan, I can imagine how new types of stories can be developed. However, from a scientific point of view, I think that these are not possible and there is only one reality.

Another scenario could be thought about the space instances of new moments. I have a very interesting and elegant alternative. There could be only one space instance. In this case, that space instance could be resetting itself between two projections and becoming a new empty space. I love this idea but I don't think that'd be true. That's because of time dilation in Special Relativity. The relation between spatial velocity and temporal speed makes me think there is something to access to get to the new space instance. Again, this is a good debate for a brain storming session.

The model proposes useful outcomes about the Quantum World which really conflicts with our common sense in many ways. Take the quantum entanglement as an example. If the waves of the entangled particle are in CES, there is no limit of speed of light, there is even no defined distance between the particles. So, there is no magic in it.

The case of entanglement inspires innovative ideas about the universe. If things can be in interaction in CES environment which are far away in ST environment than this could mean, there could be many other natural phenomena like this that we are not aware of. Maybe, the entangled particles share some complex wave components (which don't observe) that link them even if they are far away from each other in CES environment. We could speculate on many ideas like this. It is another good item to brainstorm about.

There are many such items which could be listed. DEM offers the chance to look at the universe from beyond it and defines a set of rules about how it works. If you internalize this perspective, you can find similar alternative ideas by yourself.

One last reminder about scientific scope of the book. The book contains the main definitions of the model and analyzes it vs. mainly Relativity and Quantum Mechanics. I know there are many other items (string theory, loop quantum theory, thermodynamics, details about QED etc.), I could add but I didn't. Otherwise, the book would be too complex. I think, even with this scope, it is already quite complex.

Nevertheless, the scope of the model requires to make a comment related to the String Theory and Loop Quantum Theory because they are two candidates for the theory of everything. I think, DEM has some DNA of both. Strings could be the source of Material Waves in CES distorting the flat and orthogonal N-dimensional space which is quantized according to LQT (and DEM).

That's the end of the first brainstorming session about our universe and reality. I hope, this first session has some valuable contributions to our understanding of our reality and will trigger new ideas in our minds.

Every idea, hypothesis and deduction in this book is open to critics and feedback is highly appreciated. This is definitely the first version of DEM. In time, it will evolve. It has to evolve in the search for truth.